Ecofeminism Revisited

Ecofeminism Revisited

Introduction to the Discourse

Chhaya Datar

RAWAT PUBLICATIONS
Jaipur • New Delhi • Bangalore • Hyderabad • Guwahati

ISBN 978-81-316-0399-4

Published by
Prem Rawat for **Rawat Publications**
Satyam Apts, Sector 3, Jawahar Nagar, Jaipur 302 004 (India)
Phone: 0141 265 1748 / 7006 Fax: 0141 265 1748
E-mail: info@rawatbooks.com
Website: www.rawatbooks.com

New Delhi Office
4858/24, Ansari Road, Daryaganj, New Delhi 110 002
Phone: 011 2326 3290

Also at *Bangalore, Hyderabad* and *Guwahati*

Typeset by Rawat Computers, Jaipur
Printed at Chaman Enterprises, New Delhi

For

women cultivators and women agricultural labourers
in India who provide the food security
to all of us in urban India,
and to all the grandchildren
whose future is uncertain under the
shadow of climate change.

Contents

Foreword

Given the present state of our environment and thoughtless growth models, imitative of Western models of growth, our recognition that the planet earth does not have inexhaustible resources is driving concerned scholars to examine feasible alternatives. Chhaya Datar tries to capture the different positions regarding our ecosystem of which one challenging perspective is 'ecofeminism'. Serious issues are raised in these discussions where woman and nature figure prominently. Do women have a unique relationship with nature by virtue of their special role in human reproduction? It can be argued that regenerative power of women in bringing forth a child and being devoted to its care and nurturing gives an analogy that the same nurturing quality inheres in women with regard to nature. However, this would be a 'biologistic' approach, neglecting historical and social factors. Insofar as women in many rural economies depend on gathering fuel, fodder and water for the sustenance of their households, they have a stronger stake in preserving our limited resources. Most of all we have to contend with explosive population growth which puts unrelenting demand on natural resources.

The author does not contend that we have to go back to pre-industrial modes of production but takes a middle position where our concern for improving productivity to feed a large population, and to obtain necessary raw materials, has to be tempered by the knowledge that regeneration of lost resources takes a long time. I recall an episode here. I had gone to attend a 'Women and Environment' conference at Vancouver many years ago. There was this lady who said she has given up all the appendages of industrial civilization and gone back to pristine ways. She has well to draw water from, she has field she herself ploughs, sows and harvests. She lives far away from civilization. We too have had the 'hermitage' way of life. To me these can be only brief periods of escape from the pressures of urbanization. Civilizations were built where there was human social interaction. Chhaya Datar links our present approach not only to capitalism which emphasizes growth but also to the role played by science and positivism. She argues that science has a splintered view that nature exists for us to exploit. Here, she brings in feminist criticism of science. Scientific methodology has a built-in reductionist stance. While this very characteristic has given science enormous progress in terms of our understanding about ourselves and our total dependence on nature, it also failed to establish the intimate interconnections between living and non-living matter. Underlying this is the philosophical view put forward by many evolutionary biologists in the past as well as now that we are not separate beings; we are part of a larger system and to disturb the web of relations is to court disaster. Modern biology is not reductionist the way physics might be, but propagates the idea that our life on this planet and the planet itself is an organism. Philosophers of biology like J.B.S. Haldane in the past and S. Duraiswami now explicate the notion of 'organism' and 'organismic' arrangement of nature. Stephen Gould in his latest book *Chaos* demonstrates how societies that did not take care of their environment went under. Technology has been the main villain in the peace today but Gould argues we can, given our gift of reflection, turn the clock back. Datar talks about the fact that the present pattern of growth is violent and displaces traditional communities whose lifestyle was coherent with maintaining the natural resources. The present model of growth also creates extreme inequalities. Class intrudes in any appraisal of policies and programmes.

One of the things we have to understand is that right from the beginning of settled agriculture, we have interfered with nature – only the pace was slow, spanning centuries. Historically, we have left communal modes long ago and inevitably built the accompanying hierarchy. Do we therefore go back to wilderness?

Do we all become Tarzans? It is not easy to dismantle all that development has built but we can plan a future where as Datar points out development can be more decentralized and more local. In this what role can women play? There are evidences within our own country where women have had a longer vision and objected to tree felling for timber. Have they been involved in planning? Market economy conflicts with subsistence production. Food is essential but when productivity increases to an extent where supply overshoots demand, regulation has become necessary through subsidies to compensate for low prices in Europe.

Datar's engagement in the book is philosophical and epistemological. The use of fossil fuels and the internal combustion engine has brought conveniences to individuals but has also posed threats to the environment. Yet, I wonder – while we can write passionate tomes against use of cars, aeroplanes, all of us use them. So, where does advocacy and practice meet?

All said and done, Datar has given us a rapid bird's eye view of the literature on the theme of ecology and how feminism has a connection to it. As she apologises, the literature is pre-dominantly Western mainly because we do not have enough theoretical writings. It is hoped the book will provoke discussion and debate in the classroom as well as outside. Unfortunately, our mainstream education system has an anathema to intellectual debates and intellectual engagement with critical issues. Let us hope that at least women's studies departments revive the spirit of critical inquiry. Theory informs action and our understanding.

Prof. Maithreyi Krishnaraj
Former Director, Research Centre for Women's Studies,
SNDT Women's University, Mumbai

Acknowledgement

Feminists and environmentalist have neglected ecofeminism's contribution to both theories. I hope that I was able to do justice in bringing together many strands of ecofeminism and environmentalism resonating with each other and put forth a coherent picture of ecofeminist discourse for those who are concerned with alternative development path where women's activities of reproduction and regeneration would remain at centre stage.

I am deeply indebted to Prof. Maithreyi Krishnaraj who helped me to adhere to a theoretical thread I wanted to highlight but was getting carried away with voluminous literature. Her comments on the initial draft were very pertinent and insightful which ensured that I knit my ideas tightly and make sense out of vast literature I had collected for this purpose. She provided me many latest references, which helped me to inform my arguments with new knowledges. Her readings in Buddhism were useful contribution to my understanding of the ecological perspective. These additions gave my manuscript a fresh look. If the readers are able to grasp the contours of ecofeminist ideas, the credit goes to Maithreyi Krishnaraj, and the failure for sharpness of these contours is mine.

I am in love with Maria Mies' ideas, although sometimes they sound very unrealistic, but I am always fascinated by their clarity and consistent logic. She was my Guru at Institute of Social Studies, The Hague and has remained a friend till today. I can also say that I proposed to write this book also to make her ideas visible to the feminists in India. They may discard them but at least they should take cognizance of these ideas during their discussions of relevant feminist theories in the Indian context.

I have to thank Manjiri Gupte for helping me looking for relevant books and web material for glossary. Her detailed notes were very useful for updating the material I had collected ten years earlier.

Siddharth Bhat contributed towards making my writing more readable and succinct. I owe him thanks.

Lastly, but not the least important is the contribution by the TISS for providing small pre-publication grant to update my first draft written in 2000 and provide referencing and other assistance to make the manuscript as rigorous as possible.

A big thank you to all of you,

Chhaya Datar

Introduction

The book aims at an introduction to ecofeminism as a perspective from which to understand the world around us. While primarily an academic overview, it also represents a personal odyssey. My awareness of environment perspectives was kindled through a conjunction of two important events. First, in 1986–87, I did a research project evaluating the much talked about Employment Guarantee Scheme in Maharashtra and its impact on women's employment and their empowerment. This work exposed me to the reality of dry land areas and their contrast to the verdant pockets made possible by the 'green revolution' through surface irrigation created by impounding water in costly dams. Maharashtra has many dams but they irrigate very small command area, around 8 per cent. The rest 7 per cent becomes possible through wells and bore wells.

Second, around the same time, Narmada Bachao Andolan (Save Narmada River) initiated demands for halting the construction of the Sardar Sarovar Dam, challenging the economics of dam construction in the dual light of environmental consequences to the affected areas and social consequences to the displaced tribals and peasants (see, Chapter 12).

My awareness, gained through the above mentioned two events, about the need to manage natural resources in a sustainable manner, was increased through another project I got involved with in 1992. The Government of Maharashtra started the work of community development in two hundred villages with the project aimed at the empowerment of women in decision-making and management of a drinking water scheme installed by the government. This experience exposed me to the consequences of government's shortsighted management policies towards water as a natural resource. It was obvious that the government's perspective for development was embedded within conventional technologies and centralized management systems.

These personal experiences were a part of the larger historical setting. Collapse of socialism and centralized state economies had cast complete gloom over conventional unionizing and organizing activities. Capitalism was perceived as the only and inevitable path to development (with There is No Alternative – TINA syndrome). Borrowing from the World Bank and IMF and accepting their conditionalities of 'structural adjustment' has been the path chosen by the Indian government since 1991. Primarily it involves reallocation of public finance, reducing state's fiscal responsibility, e.g., instead of free healthcare, charging 'user fees'. Liberalization, privatization and globalization (LPG) became key policies and although the negative impact was increasingly felt, these were accepted as an inevitable consequence of a bitter pill for a brighter future.

Economic changes through LPG policies increased requirement of cash in the rural economy, which aggravated the impetus to the small landholders and peasants to move away from subsistence farming and either keep the land fallow or give it to the large landholders for cultivating cash crops for the market. In Punjab contract farming was introduced as a new way of bondage, replacing earlier feudal bondage, where tenant peasant could not leave land and achieve freedom to look for a job in the factory. In the new contract system the agro-companies promised to buy the produce provided that the farmers agree to produce the crop by using inputs supplied by the company. Although it helped to improve productivity of land in the short term, in the long term, the impact of the excessive chemical fertilizers and pesticides was felt in terms of deteriorating soil conditions. Monoculture displaced the diversity of traditional patterns of cropping, and it was supported by using bore wells which went deeper and deeper, further depleting an underground water table that was not easily rechargeable. Awareness of these events impelled me to search for a cohesive theory which would explain not only women's subordination in the process of accumulation of capital but also its relationship with environmental degradation. I was also looking for a theoretical framework within which to develop a new vision for a society where the ideals of equity, justice and sustainability of natural resources could be realized.

Thus emerged the present book of introduction to the discourse of ecofeminism. It is not an attempt for developing my own theory. It is also not an attempt to read and summarize the discourse of ecofeminism, introducing all the streams and facets of this discourse or trying to balance the space between representation given to each stream. Instead, I have tried here to synthesize many ideas, theories and strategies which seem compatible with each other or at least echo similar concerns and thus contribute to enrich and deepen the understanding of the relationship between women, men and nature, which ecofeminism is trying to do. Although I have made sincere efforts to touch upon all the major dimensions of ecofeminism as they are discussed, debated and theorized, these contributions ultimately appear as the building blocks of a mega-discourse as it should develop and become more and more coherent.

I still believe in the famous Marxian maxim that "it is not enough to understand the world, but the point is how to change it". In service to this goal, I have collected and collated the ideas, views and theories of authors from several diverse and sometimes disparate discourses which constitute environmental and feminist literature. These discourses do not necessarily acknowledge each other and may not agree to all the intricacies and details of each others' point of view. I have used my personal judgement to cull out from their writings that which would contribute to the perspective that might guide many of us through our daily lives and actions.

Structure of the Presentation

The first and introductory chapter on ecopolitics tries to show the paradigmic contrast between the often known strands of feminism such as liberal, Marxist/socialist, radical socialism and ecofeminism. For the sake of convenience I would like to call first three strands as either mainstream or equity feminism, because without being critical of the present development discourse based on surplus accumulation processes of capitalist economies, they narrow down their concern to gender equity within the framework of national economies. Ecofeminism, on the other hand, envisions the gender equity in an overarching framework of abolishing hierarchies and power relations between humans and non-humans, between the First World and the Third World and also between men and women. The later part presents emergence of ecofeminism with its different strands. This information is important because mainstream feminists collapse all strands with cultural ecofeminism and thus accuse ecofeminism of essentializing sexual identity of both women and men. Romanticizing feminity and mothering is another accusation levelled against the ecofeminist discourse in general. For this book I have chosen two main streams of ecofeminism: socialist feminist represented by Carolyn Merchant and subsistence perspective strand represented by Maria Mies, Veronika Bennholdt-Thomsen and Claudia Von Werlhof. There are substantive presentations from both on diverse themes handled by them in the context of feminism, environmentalism and development (capitalism and industrialism). Vandana Shiva also echoes similar views and her writings are also often quoted.

In the second chapter, I present the brief history of various environmental movements both in the West and in India where women were engaged prominently. These movements exposed the diverse dimensions of ecological destruction, induced by development policies. There are three annexures attached out of which two present the case studies of two well-known struggles waged by women in an early period of the emergence of environmental movements. Greenham Common struggle was waged against setting up a site of nuclear missiles in UK. Chipko movement in Himalayas, in India, was against the felling of trees by timber merchants in connivance with the forest department. The third annexure gives summary of the environmental movements worldwide. The chapter presents wide range of issues which were raised during these struggles and also shows contrast these movements had with traditional movements of the working class. Here marginalized people such as women, peasants and slum dwellers came together along with the middle class and intellectuals. Environmental movements are considered as a part of the New Social Movements which questioned the slogan of "catching up with the West".

The third chapter deals exclusively with impact of environmental degradation on women in India. It brings together readings which focus on women in the rural areas. There are various reasons for this choice, the most significant being that environmental movements have not thus far highlighted the demands of affected women in their protest against the destruction of natural resources. Another reason to give special attention to the concerns of rural women is to address an imbalance in the way environmental problems are studied: the impact of industrialization is recognized (air pollution, garbage problems, congestion, slums of the urban settlements) but very little awareness exists regarding the rural women whose daily livelihood depends upon unmediated natural resources.

Chapter four deals with the critique of industrialism stating that institutions created by the industrial revolution are responsible for the ecological crisis. The nexus between market, technology and environmental economics creates inevitably centralization of power and also money, which promotes reckless use of natural resources without caring for the exigencies of local ecosystems. The critique also points out the marginalization of women in this process. Gandhian critique of policies that questions blindly following of the West has also been given some space. Decentralization of production systems and communities so that the people can participate in the decision-making with utmost concern for the ecological sustainability of their natural recourses has been advocated not only by Gandhi, an idealist, but also by many other economists and social thinkers. An excellent illustration is the policy of SEZ pushing industrialization at the cost of agriculture and the agricultural communities. Loss of land under SEZ implies loss of livelihood, exposing agricultural communities to risks they cannot bear. There is an annexure provided on SEZ.

The fifth chapter looks at changing relations between man, woman and nature historically. It also means periodization of history from the perspective of changing resource use, i.e., ecological history. Tracing of emergence of sexual division of labour and dominance of man over woman in history is the important step in locating the relationship between man, woman and nature. Similar kind of hierarchies in the use of natural resources gets expressed in terms of castes and class in the Indian situation, which have a bearing for development of technologies. Technologies determine relationship of man with nature. These theories are rarely scrutinized by the critics of ecofeminism. I strongly feel that these theories, if synthesized properly, could conform to a materialist conception of history, as the core of ecofeminist thought. Ecological thoughts and ecofeminism complement each other and, with their separate expertise can fill the gaps to create a workable platform for change.

Chapter six explores how through history (Western and Indian) the representation of women and nature has undergone changes. There is a need to go behind some of the images where identification of women and nature is forged. These images lie in the prehistory – a period of Mother Goddesses. Many of these images and symbols are recently being unearthed, through archaeological excavations and being interpreted by cultural anthropologists through studies of folk culture and rituals. These iconic images showing a link between women and nature exert influence till today. Residues of this ancient period of mother pre-eminence are definitely found in many rituals in different religions the world over. These old representations sometimes are used by ecofeminists as forms of resistance. Many anthropologists studied these historical representations to investigate how women and nature were slowly relegated to passivity in the newly emerging rationalist tradition.

Chapter seven focuses on the epistemological aspects of the process of industrialization, which helped build the paradigm of the Enlightenment period. Many feminist scientists as well as social scientists studied the history of scientific philosophy and found out how 'reason' and rationality were promoted to characterize 'humanness' relegating 'sensuous experiences' to a lower level of existence. The critical role played by Bacon is acknowledged by both socialist feminists as well as ecofeminists, in assigning a superior role to man for establishing mastery over nature. Cartecian division between 'mind' and 'body' further helped in identifying woman with nature. She was 'body' while man was assigned superior realm of abstract thought. Ecofeminism holds that this dualist rationalist tradition made women invisible. And, this invisibility was not restricted to women alone, but was endemic to the scientific view of the processes of nature, of which women were perceived as a part. The positivist logic used in the rationalist tradition promotes subject-object dichotomy and fails to understand the complex web of ecological processes. It cannot arrest destruction. This discussion reveals the nexus of technology, market and modernism, or as Maria Mies calls it capitalist patriarchy and its progressively ascending power and control over natural resources.

Chapter eight talks of ethics. The environmental movement has also widened the scope for discussion about the ethics of domination of man by man to encompass what constitutes domination of nature by human beings. Ultimately, the question of domination focuses upon the philosophical question of how to conceptualize the human 'self', which determines its relationship with 'others' – both humans and non-humans. Ecofeminists are deeply concerned about these philosophical ideas. An intricate debate has been triggered by the feminists on a particular issue of eating animals/meat. The passionate ecofeminist view is that because we reduce animals as culturally edible, we start raising them for factory production. These ecofeminists believe that there is a difference in tribals killing animals because of threat or for eating as food and raising animals in the factory for eating. I found a pertinent question raised by Robert Goodin as compatible with ecofeminist ideas: "how much intervention of nature is tolerable?" His argument enriches our understanding in terms of what constitutes an ethical role while we lead our day-to-day life, dependent as it is upon natural resources.

Chapter nine discusses frontier technologies such as genetic engineering and reproductive technologies, where the ethical issues involved demand urgent attention. Only ecofeminists have paid critical attention to the dangerous implications and consequences of these technologies whereas rationalist feminists, i.e., liberal and Marxist, seem to look at them as means of further emancipation, matter of choice and self-determination. They exclude nature and the ethics of our relation to nature, necessary for approaching complex web of ecological processes. Information on FINNRAGE in the annexure provides the perspective for opposition to the reproductive technologies. There is also an annexure on farmers, who cannot afford the new technology, resisting in Karnataka for introduction of genetically engineered seeds. Given their concern for biodiversity and self-reliance over seed companies, they do not appreciate these technologies.

After mentioning all the major dimensions of ecofeminism, I feel an examination of efforts to propose alternative technological paradigms, especially those that promise to realize the vision of a dispersed and decentralized society becomes inevitable. Much of the literature on alternatives is mainly visionary except some which propose strategies, but not in the usual sense of policies to be adopted by the governments. They have anarchist elements and depend upon the rebellion from the grassroots. There is only one argument which is very well worked out and even believes that it would require central agency to universalize certain principles of equality in the use of natural resources after some successful experiments have been proven to be workable. The literature presented in chapter ten is limited and there is an ongoing need to identify more material on this theme.

I discuss the criticism of ecofeminism in chapter eleven, so that readers get to know both sides – liberal and socialist feminist perspectives vis-à-vis ecofeminism as a perspective. One of the reasons I chose this style of writing, i.e., collating diverse writings which covers both ecofeminism and environmentalism, is that I feel at present criticism against ecofeminism is based on fragments of ecofeminist ideas and hence does not do justice to the world view within which the perspective is developed. Someone like me is eligible because I claim no new thought or theoretical contribution but feel the need to put together bits and pieces from the body of knowledge produced since the sixties centred around the issue of relationships between humans and nature and between man and woman. I am hopeful that it will help in developing a coherent perspective as a guide to action. The experience has been very enthralling for me personally and I hope it stimulates the readers too. It was a paradigm shift for me from the 'known' mainstream, equity feminism to ecofeminism. The new paradigm changed for me the fundamental image of a subject matter of women and nature. It gave me new purpose, new orientation, new methods of enquiry and also new strategies for resolving the problems.

Chapter twelve is in fact an annexure to illustrate that some elements of equality and sustainability have been already operative at the grassroots and the visions developed by ecofeminists and others are not abstract and idealist. I have chosen three sectors, viz., forest, land and water, and presented some case studies of NGOs which are trying to undertake experiments and put their ideas and visions of new decentralized and participatory society into practice.

The last chapter 'Conclusion' brings together my arguments in brief.

I would like to note here my observation that the feminist scholarly field has sometimes allowed itself to become polemical and/or egoistic, which makes authors reject the ideas put forward by others with the easy tool of textual misinterpretation. It also appears that theoretical discussions are getting more and more restricted to the use of certain language, metaphors and conceptual discourses. Such a discourse, if it excludes those who are concerned more with strategy, practice and action, needs to be questioned. Some feminists fight these intellectual battles with such animosity and bitterness that they seem to assume that these fights play a proxy role to political action. One also has to acknowledge the likelihood that there are territorial wars going on in the commercial publishing world and authors need to maintain hegemony over certain discourses. I have used Indian material and also Western/Northern. Earlier I had thought of privileging Indian material over the Western material in each chapter by sequencing it first. But I soon realized that the Western material is overwhelming and also covers multidimensions. Hence, it is more practical to cover Western material first and then document the Indian material. I hope the readers will accept this deliberately action-oriented presentation.

1

Feminist Ecopolitics

> The denial of material, individual and social histories goes hand in hand with the hope that it will give women at least an access to the technologically and patriarchally defined realm of men. This realm is seen as the 'realm of freedom' and of 'culture'. The old dream of all oppressed people – to move up into the house of their masters instead of breaking down this house – is also many women's dream. In politics as well as in women's studies some women have indeed been able to move into that house. But they were accepted by the 'Male Stream' (O'Brian, 1989) only after their denial of the origins, the separation from the women's movement and reacademisation of Women's Studies had taken place. (Mies and Bennholdt-Thomsen, 1999)

While being taught feminism, students are primarily introduced to a specific taxonomy – Liberal Feminism, Socialist/Marxist Feminism and Radical/Cultural Feminism. Ecofeminism usually does not figure on this list and is brushed aside as a combination of environmentalism (of the deep ecology type) and the worship of Mother Goddesses (cultural ecofeminism). Feminism associates itself with 'liberation', 'emancipation' or 'self-determination' and 'modernism' distinct from the 'traditional images' of women. These images (traceable to the advent of civilization) are regarded as being embedded in a hierarchically organized 'sexual division of labour'.

Most strands of feminism uphold, in varying degrees, the modernist dichotomy between nature and culture. Simone de Beauvoir, in her book *Second Sex*, points out that this distinction equates women with nature (characterized by their biological composition) and men with culture (characterized by their 'risk-taking' behaviour). Liberal and Marxist feminists argue that the traditional notion of a connection between women and nature is a relic of patriarchy – an instrument of oppression – which should be allowed to wither away. For them, ecofeminism smacks of essentialism (biological determinism).

Despite such criticism, one needs to acknowledge the fact that exploring ecofeminist arguments rising from a material base (social, historical, dialectical) creates support in favour of alternative development models (as opposed to market-oriented capitalist ones). Poor women often find a potential for liberation within such models. It also provides a better understanding of movements like the Narmada Bachao Andolan, opposition to SEZ etc., which strongly emphasize Third World women's concerns for food security and as such their vested interest in the preservation of ecological bases for the survival of their communities. Concepts like 'decentralized communities', 'subsistence production' etc., need to be understood against a theoretical background which justifies the need to start thinking about alternative development models.

Feminist theorization does not generally include the category 'nature'. Within feminist discourse, the issues of environment and development addressed by developmental feminists come closest to a concern for nature. Sandra Harding (1998: 113) mentions a platform called, "Women, the Environment and Alternatives to Development (WEAD)" and refers to the rich resources produced by people like Wendy Hartcourt under the title "Negotiating Positions in the Sustainable Development Debate: Situating the Feminist Perspective". According to her, these resources prove that there has to be a gendered standpoint on nature. Even though postcolonial discussions argue for the role of local knowledges in the process of development, they do not point out the geographical location of women within Southern countries. The hierarchical location of the 'other', i.e., women, is not taken into account. However, Harding's exploration of Feminist Development literature does not trace the need for a paradigm shift.

She merely admires the argument of mainstream feminists that 'economic growth' needs to be attuned to cultural specificities of communities in the South, particularly the women in these communities who occupy a lower rung.

Glimpses of Ecofeminism: By Way of Beginning

There are various contradictory strands within ecofeminism. However, each strand firmly believes that nature (along with gender, race and class) has to become a category of analysis to fully understand women's oppression. They assert that the mutual inferiorization of women and nature continues to fuel a denial of women's activity and the whole sphere of reproduction. Ecofeminism has contributed enormously to activist struggle and theorizing. Over the last two decades, it has brought the links between domination over nature and women's oppression into general awareness. Still, as Val Plumwood laments, it has been stereotyped as being theoretically weak and as doubtfully liberated. Furthermore, it is frowned upon as being exclusively linked to cultural feminism (Plumwood, 1993: 1).

Plumwood is one of the few ecofeminists who tries to develop the theoretical underpinnings of environmental degradation along with the continued devaluation of women's reproductive activities. This is achieved by questioning the 'dualistic rationality' of mainstream scientific philosophy. Second-wave feminism analyses the phallocentric development of the world (penis – symbol of man's power). Plumwood takes this argument a step further by attempting to understand the relationship between phallocentrism and anthropocentrism (human species as the centre of the ecosphere). She points out the need to alter the way in which we conceptualize the world around us.

Plumwood has also tried to explore the differences and continuity between feminism and ecofeminism (ibid.). One evident difference is the suspicion with which feminists view the idea of a feminine connection with nature. To this school, this association seems regressive and insulting as it summons images of women as passive, reproductive animals who are limited to their bodies and an unreflecting experiencing of life. A basic difference between the two schools of thought, as she sees it, is that ecofeminism revolves around an uncritical acceptance of the concept of 'equality'.

During 1960s and 70s feminists attempted to fit women into the masculine model of humanity and culture, science and technology from which women had been excluded. These areas involved highly valued masculine traits of objectivity, abstractness, rationality and suppression of emotionality. These fields also strongly exhibited masculine virtues like transcendence of, control over and struggle with nature. According to Plumwood, ecofeminism believes that women cannot be accommodated in a cultural ideal, which has defined itself in opposition to feminine traits (reproduction and caring). She also asserts that the 'problem of equality' cannot be resolved by adopting the reversal strategy extolled by cultural feminists (which involves attaching positive value to the feminine and natural).

Plumwood asserts that the problem lies in the epistemic concept of dualism embedded in the masculine knowledge system, which constructs 'difference' using a hierarchical logic. The more valued side (like masculinity) is constructed as superior, alien to and of a different nature to the 'lower order' (like femininity) which is perceived as lacking some qualities. This dualism views both women and men as damaged or distorted. With an ecofeminist position, it is possible to recognize 'human' identity as continuous with nature and not alien to it. All ecofeminist positions reject the assumed inferiority of women and nature as opposed to the superiority of reason, humanity and culture (ibid.).

A second set of ecofeminists (comprising Vandana Shiva, Maria Mies and her German friends Claudia Von Werlhof and Veronika Bennholdt-Thomsen) seeks to build a theory from the practical experiences of Third World women at the grassroots level. These experiences are two-fold: (a) the impact of marginalization caused by the depletion of natural resources along with dislocation due to development projects, and (b) alternative strategies of survival as a resistance to development. These ecofeminists focus on a materialist understanding of society to alter the current exploitative paradigm which is sustained and legitimized by patriarchy.

Mies and Shiva also criticize the dominant concept of rationality. They reject the epistemological paradigm based on Cartesian hubris (dominance of men over nature and women). This rejection can be traced to the experiences of women's movements which resisted patriarchal violence, militarism, nuclear and gene technology. They cite the influence of T.S. Kuhn's publication of *The Structure of Scientific Revolution* (1962) on scientists like Merchant (1980), Keller (1985), Chodorow (1978) and Gilligan (1982). It asserts that scientific theories are influenced by the world view prevailing at that point of time. Within the scientific community, the development of a new paradigm and setting up of new research questions is influenced by social factors apart from empirical evidences and theoretical necessity (internal logic).

In their book *The Subsistence Perspective* (1999), Maria Mies and Veronika Bennholdt-Thomsen describe their discontent with the present women's movement. They claim that the beginning of the new women's movement did liberate women. In the seventies, 'women's liberation', 'freedom from exploitation and oppression from sexism and violence' and 'fighting capitalist patriarchy' were key concepts and goals. The discovery of their commonality gave women a strength they had not known before. They learnt to love women just because they were members of 'second sex' (de Beauvoir, 1974). However, this empowering new sisterhood did not encompass the entire spectrum of female existence worldwide. The 'differences' (of class, race, culture, ethnicity, religion, sexual orientation, age etc.) have became more important than the commonalities embracing all women. The authors are critical of the demands of women's movements in many countries especially those which are supported by NGOs implementing the 'Out of Margins' and 'Into the Mainstream' paradigm. These movements call for 'empowerment' of women instead of their liberation. While this strategy promises women money, freedom and equality, it mystifies the fact that within capitalist patriarchy only a minority of women will be able to achieve these goals. Another criticism levelled against the women in the North is that their assertion of 'human rights' is basically an attempt to share money and power equally with men (Mies and Bennholdt-Thomsen, 1999: 182–206).

The authors notice the same trend in postmodern feminist positions (1980 onwards) which reject motherhood and social history (body politics – the struggle against violence). Postmodernism has engulfed contemporary feminist thinking based on a new idealism, a need for self-construction and an emphasis on analysing textual and virtual reality. The authors observe that feminists are increasingly vocalizing an intense dislike against 'body feminine' and 'motherhood', i.e., denying a biological basis for woman's identity. They also claim that the arguments about the essentialist nature of ecofeminism have their roots in a denial of our origins as 'of woman born' – of real mothers and their genealogy. Thus gene and reproductive technologies become the means to emancipate women from the supposed 'wilderness' of their own body. Postmodernist feminists strongly resist prioritizing any single feminist issue for struggle as for them everything stands on a level plane.

In contrast to the current 'mainstreamization' position, Mies and Bennholdt-Thomsen state their theoretical position in the following way:

1. The main problem for women the world over is not difference and/or identity. Rather, it is exploitation, oppression, violence and colonialism. Capitalist patriarchy, which cannot tolerate equal valuation of all, has managed to ideologically transform diversities into antagonisms. Hence, every 'other' becomes an enemy – a competitor – within a world of scarcity. We want to create a world in which biological and cultural diversity is maintained and celebrated.

2. We remain, on this earth, connected to all other creatures. We do not expect freedom, wealth, happiness and a 'good

life' to come from some kind of transcendence – beyond the realm of necessity – be it the transcendence of religion, of money or of postmodern virtual reality. We continue to celebrate matter as the basis of a good life. We reject dualism that divides matter from spirit, devaluing the former while idealizing the latter.

3. Our empowerment is not based on technocratic domination over other creatures or other human beings, nor on participation in the capitalist and patriarchal power structures, but on self-reliance and autonomy, mutuality, self-organization, self-provisioning, local and global networks and subsistence relations instead of profit relations.

4. We know that we are historical beings. We know that without the knowledge of and the respect for female genealogies, women will not be able to overcome patriarchy either individually or socially. We want to reclaim and reconnect ourselves with our daughters.

5. We also want to reclaim and connect ourselves with our sons. We will not accept that our sons have no other perspective than that offered by macho, militarist global capitalism (ibid.: 202–03).

Mies and Bennholdt-Thomsen illustrate the dynamics of the new international feminist community through the celebration of 'Woman's Day on Food' (organized on 15th November, 1996 during the FAO World Food Summit in Rome) which was jointly organized by women of the South and the North. They advocated that 'food security' must remain in the hands of women. Local and regional self-sufficiency should replace the globalized food industry of transnational companies (TNCs). On the same day Vandana Shiva and others initiated a new international network called 'Diverse Women for Diversity' (Mies, 1996).

Two broad conceptual positions are discernible in these arguments – Marxist-liberal (developmental feminist) and ecofeminist – from which two different strategies for women's liberation emanate. Critics are engaged in a heated debate accusing both sides of methodological distortions.

A Brief History of Ecofeminist Ideas in the West

According to Diamond and Orenstein (1990: ix), the development of ecofeminism took four different paths in America. Each path has a different background and history. Even though they lay emphasis on different issues, their common concern is to prevent the destruction of nature.

The first path rejected the Marxist assertion that domination is based solely on money and class. Radical feminists shifted their attention towards ecofeminism after realizing the inadequacies of orthodox Marxist theory (which ignores both nature and women). Another source of the radical/cultural feminist dominance theory was the work of cultural historians (like Riane Eisler, Susan Griffin and Elizabeth Dodson Gray) who explored the roots of patriarchy.

The second path which led to ecofeminism was the exposure to nature-based religion (usually based on goddesses). In the mid-1970s, many radical/cultural feminists made the exhilarating discovery (through historical and archaeological sources) of a religion which honoured the female. The essence of their attraction was a wholesome and healing cosmology supported by the Divine as immanent in and around us. The early period inspired art, music and poetry. In the words of Spretnak, "In the beginning ecology was not on our minds, but since we moved out of that period into activism, ecology has never left our minds" (Spretnak, 1990: 11).

The third path into ecofeminism came from environmentalism. This path is generally credited to the publication of Rachel Carson's *Silent Spring*[1] (1962). Carson fought against pollution and the degradation of nature. Although she did not call herself a feminist, her text prefigured the powerful USA environmental movement which culminated in the nationwide Earth Day of 1970 – an event in which many feminists participated. In 1974, the Conference on Women and Environment at the University of

California, Berkeley spurred ecofeminism into a vigorous movement. Sandra Marburg, Lisa Watson, Ynestra King, Grace Paley, Anna Gyorgy and Christina Rawley can be cited to mention a few significant figures (Diamond and Orenstein, 1990). Many of them got together again in Amherst, Massachusetts to discuss issues related to environmental destruction and its consequences on women when the Three-Mile Island catastrophe[2] took place.

The fourth path is attributed to women with careers in public policy, science and technology, public interest environmental organizations or environmental studies programmes in universities. Their initial contact with feminism was through liberal feminist attention on how and why a woman's progress in her career was impeded. Concomitantly, they were concerned about issues like the rich-poor divide, the associated polarization between the First World and the Third World and within the Third World itself. Later (through contact with ecofeminism) the trajectories of their career acquired a different direction altogether.

Similarly, many people discovered ecofeminism and deep ecology through their involvement with green politics. Women realized that most ecocentric world views did not recognize their role in the reproduction of humankind – work which involves the nurturance of inner and outer nature. Ecofeminists observed that ecocentric philosophies (which rejected the modernization paradigm) could not accommodate activities (such as daily and intergenerational reproduction of women) in their conceptualization of nature. Moreover, they were willing to learn sustainability norms women had set up and practised for centuries. Mies and her colleagues have strongly argued in favour of this trend. They preach the need to create a culture which will highlight the interconnectedness of living systems and the interdependence between human beings and nature. The similar ideas are voiced in Buddhist philosophy (Keown, 1996: 51). They reject the idea of achieving emancipation by competing with men in the mainstream economy. They are also strongly against the modernization paradigm, which they find anthropo (human)- and andro (male)-centric.

Mies and her colleagues are criticized for their extreme views on technological development because many technologies (like fertility control) have benefited women globally and have contributed towards nurturing modern cultural values such as self-determination. Another critique is their silence on population control which is a major burden on the carrying capacity of earth.

Diverse Trends Among the Ecofeminists

In her book *Earthcare,* Carolyn Merchant maps the entire gamut of feminist approaches which display a concern for ecological issues (Merchant, 1995: 9–18). Merchant wants to be inclusive and hence also discusses inadequately developed approaches. By combining socialist tenets with ecology principles, Merchant develops a new perspective which she hopes will be acceptable to socialist feminism. However, both socialist feminists and postmodernists brush aside ecofeminism on grounds of essentialism.

Liberal Ecofeminism

Liberal feminism characterized the history of feminism from its beginnings in the 17th century till the 1960s. It is rooted in liberalism (a theory of human nature which views humans as individual rational agents who maximize their own self-interest, a political theory which accepts the scientific analysis that nature is composed of atoms moved by external forces), which regards capitalism as the optimal economic structure for human progress. It also accepts the egocentric ethic that an optimal society results when each individual maximizes their own productive potential. Thus, what is good for each individual is good for society as a whole. Historically, liberal feminists have argued that women do not differ from men as rational agents. Exclusion from educational and economic opportunities has prevented them from realizing their own creative potential in all spheres of human life. Environmental problems are caused by excessively rapid development of natural resources and failure to regulate pesticides and other environmental pollutants. Given an opportunity, women can transcend the social stigma of their biology and join men in the cultural project of environmental conservation.

Cultural Ecofeminism

Cultural feminism is a response to the perception that women and nature have been mutually associated and devalued in Western culture. Historically and across cultures, women have been perceived as closer to nature because of their physiology, social roles and psychology. Physiologically, women bring forth life, undergoing the pleasure, pains and stigmas attached to menstruation, pregnancy, childbirth and nursing. Socially, child rearing and domestic chores have kept married women closer to the hearth and out of the workplace. Psychologically, women have been assigned greater emotional capacities with greater ties to the particular, personal and present. Men are viewed as more rational and objective with a greater capacity for abstract thinking. Their psychology leaves them freer to travel, hunt, conduct warfare and engage in public affairs.

For cultural feminists, human nature is grounded in human biology. The perceived connection between women and biological reproduction is cited as a source of women's empowerment and ecological activism. This form of ecofeminism has largely focused on the sphere of consciousness in relation to nature-spirituality, goddess worship, witchcraft and the celebration of women's bodies. It is frequently accompanied by social activism such as anti-nuclear or anti-pornography protests.

A lot of populist ecological activism by these women is implicitly motivated by the connection between women's reproductive biology (nature) and the male-designed and produced technologies (culture). Other feminists employ cultural ecofeminism to represent ecofeminism and accuse it of essentializing the sexual identity of both women and men. Critiques allege that it lacks a serious analysis of capitalism and does not explain why it dominates nature. Thus, it cannot provide an effective strategy for change.

Social Ecofeminism

Social ecofeminism accepts the basic tenet of social ecology that the idea of dominating nature stems from the domination of humans by humans. It aims to liberate women by overturning economic and social hierarchies. These structures plunge all aspects of life into a market society. It envisions a society of decentralized communities which transcend the public-private dichotomy essential to capitalist production and the bureaucratic state. Women emerge as free participants in public life and in local municipal workplaces. Social ecofeminists criticize cultural feminism on the ground that it rejects rationality and (by worshipping Mother Goddesses) biologizes and essentializes the caretaking and nurturing traits assigned by patriarchy to women.

Socialist Ecofeminism

Merchant accepts socialist ecofeminism and hopes that it will augment a feminist transformation of socialist ecology thus making the category of reproduction (rather than production) the central concept of a just, sustainable world. Like Marxist feminism, it assumes that non-human nature is the material basis of all life and that food, clothing, shelter and energy are essential to the maintenance of human survival. Nature and human nature are socially and historically constructed over time and transformed through human praxis. Merchant believes that socialist ecofeminists assume that nature is an active subject (not a passive object to be dominated) and that humans must develop sustainable relations with it. This goes beyond cultural feminism in offering a critique of capitalist patriarchy which focuses on the dialectical relationships between production and reproduction and between production and ecology.

According to Merchant, a socialist ecofeminist perspective offers a standpoint from which to analyse social and ecological transformations. It suggests social actions which will lead to sustainability of life and a just society. Socialist ecofeminism claims that capitalism's logic precludes sustainability. It is premised on economic growth and competition rendering both nature and waste as externalities[3] to profit maximization. On the other hand, the logic of socialism is based on the fulfilment of people's needs. Since growth is not essential to the economy, socialism creates a possibility of sustainable relations with nature. Even though state socialism has been based on growth-oriented industrialization and has resulted in the pollution of external nature, new forms of socialist ecology could bring human production and reproduction into balance with nature's production and reproduction. Nature's economy and human economy could enter into a partnership.

Merchant focuses on the reproduction of life which, for her, is both biological and social. Like Irene Diamond, she is concerned about the implications of 'population control' on Third World women. Merchant particularly notes recent family planning techniques (including injectable contraceptives, vaccines against pregnancy, hormonal implants which are often banned in western nations as unsafe) whose indiscriminate field testing reduces women of the South to mindless objects. Prenatal sex identification raises the fear of a potential "death of the female sex". Feminists decry using women's bodies as production machines to test contraceptives, for in vitro fertilization experiments, and to produce babies for organ transplants.

Merchant defines reproductive freedom according to the demands of Third World women's groups (i.e., equal access to employment, equal pay, child care centres, and social security) so that women can choose whether to have children and in what number. Merchant argues that socialist ecofeminists are concerned with the problems faced by Third World women since they themselves are historically traceable to the colonial relations between the First and Third Worlds. At the same time, socialist ecofeminist agree that 'reproductive choice' cannot be an individual choice, but has to be a located in a social realm of public policy, which has to be facilitating to enable women to make balance choices.

Merchant asserts that from the 17th century onwards, European colonization initiated a colonial ecological revolution in which a complex combination of European animals, plants, pathogens and people disrupted native people's life and processes. The late 18th century ushered a capitalist ecological revolution in the northern hemisphere which accelerated the extraction of cash crops and resources from the southern hemisphere. Third World people were pushed to marginal lands to fill the pockets of the First and Third World elite. In the 20th century, northern industrial technologies and policies were exported to the South in the form of development projects. Green revolution agriculture (seeds, fertilizers, pesticides, dams, irrigation equipment and tractors), plantation forestry (fast growing non-indigenous species, herbicides, chip harvesters and mills), capitalist ranching (land conservation, imported grasses, fertilizers and factory farms) and reproductive technologies (potentially harmful contraceptive drugs, sterilization and bottle feeding) have further disrupted native ecologies and peoples. (I feel that she unduly favours socialist feminism even though neither liberal nor socialist feminism has displayed any marked concern for environmental issues).

Relevance of Ecofeminism in India

This knowledge is relevant to ecofeminists in India. During the last two decades, women have participated prominently in environmental struggles at the grassroots level. The ecofeminist movement has been a convergence of feminist and environmental movements. Its empirical and experiential characteristics play an important part in the Third World which suffers from the 'catching up syndrome' mooted by the developed world. In the Indian context, ecofeminism plays a role comparable to that of postcolonial theory as it analyses the imprints of colonial legacy on the land, the intellectual elite and institutional structures.

Ecofeminism challenges the concept of development and Enlightenment and thus finds itself closer to a Gandhian legacy than the Nehruvian passion for modernization. Some of these women come from a Marxist/socialist background and carry the experience of organizing workers and peasants. Some were a part of Jayaprakash Narayan's movement for Sampoorn Kranti (Total

Revolution) which in 1973–74 threatened the Congress government led by Indira Gandhi (who responded by declaring the 1975–77 Emergency).[4] The movement was partly a response to the disillusionment caused by the economic crisis, inflation and price rise suffered by the Indian economy under the economic policy of a mixed economy with a socialist ideological rhetoric.

Many of these women are participating in larger environmental movements and trying to push a women's agenda on their charter of demands. In an attempt to find a new perspective/framework for their actions and positions, their ideological battles are fought under the rubric of Gandhism vs Nehruism.[5] Vandana Shiva worked with Gandhian activists when she took part in the Chipko Andolan in Garhwal. K.R. Datye (who is trying to develop an alternative technological paradigm) says that his vision and strategies are guided by neo-Gandhism, i.e., advanced science (and not by reductionist science).

Even though this presentation uses more Western material than Indian material, it addresses people who are active in environmental movements, academicians and white-collar workers who are concerned with developing a cohesive standpoint on issues concerning women and development and women and the environment. The Indian material is more easily accessible and (at present) more descriptive. The Western material echoes the same concerns and attempts at theorization. An introduction to these theories will help develop a more enriched ecofeminist theory in the light of its practice in India.

2

Environmental Movements

Questioning Development Processes and Establishing Ecological Principles

> In an industrial society, as the impact of manufacturing activities on the environment gets externalized, women too become alienated from nature. Reactions begin when the cumulative effect is felt. (Merchant, 1995)

Ecofeminist theories are rooted in the actions of people who are affected by environmental degradation and are influenced by activists concerned with creating a humane society. The information presented here is based on the collated accounts of a few authors and does not claim to be comprehensive. It intends to show a wide range of issues put forward by various groups resisting mainstream development. The problems taken up in America have been discussed using Carolyn Merchant's (1995) views. In the US, the major environmental concerns seem to be pollution, biological degradation and climate change. Indian environmental movements are covered using two presentations – the first by Janki Andharia and Chandan Sengupta (1998) and the second by Madhav Gadgil and Ramchandra Guha (1998). Here, the primary concerns revolve around the loss of livelihood (due to degradation and displacement) and survival. Environmental movements have had a widespread impact particularly among the middle classes in both the North and the South. Annexure 1 covers the brief history of environmentalism across the world. Annexure 2 and 3 elaborate on the Greenham Common Movement (UK) and the Chipko Movement (India) respectively.

Environmental Movements in America with a Focus on Women

Before focusing on women's participation in environmental movements, one should have a brief understanding of mainstream environmental movements. Environmental movements are regarded as a part of New Social Movements (traced to the student revolts in Western Europe and US during 1968). These 'new' movements are in contrast to traditional workers' movements, which questioned modernism and economic progress by focusing on the cost of social transformation. Environmental conflicts were added to this agenda later. They were supported by feminists, minorities, young people and sometimes intellectuals. Even though their characteristics differ (in finding new approaches to environmental conflicts, creating new visions etc.), they share common interests, values, experiences and a feeling of belongingness to certain places and environments. They use unusual patterns of public protests such as civil disobedience and collective action.

I summarize below some of the major ideas and arguments of Carolyn Merchant in *Earthcare* (1995: 146). The word 'ecology' is derived from the Greek word "oikos" meaning house. Hence, ecology means the science of the household. Historically, the connection between the earth and the house was maintained by women. Productive and reproductive activities took place inside and around the house thus highlighting its relationship with the surrounding earth. In an industrial society, the impact of these activities on the environment gets externalized thereby alienating women from nature. Reaction only begins when the cumulative effect is felt.

Merchant traces the relationship between American environmental movements and women to the colonial period when indigenous religious ethos (centred on the animate vegetative world) was discarded. The 'capitalist ecological revolution' took place after the American Revolution, i.e., about 1860. The capitalist revolution demanded an economy of increased human labour and land management, which in turn required a legitimization of mechanistic science. It split human consciousness into a disembodied analytical mind and a romantic emotional sensibility. Merchant believes that ecological revolutions are processes through which different societies change their relationship with nature. They occur because of tensions between production and ecology and between production and reproduction. The results are new constructions of nature (materially and in the human consciousness). Women's changing roles are linked to this new economy both in material life and in representational systems.

As women's farm (outdoor) and textile production (indoor) declined, their reproductive roles were elevated. They became moral mothers and middle class housewives and began to influence consciousness about nature at home. Later, they functioned more publicly through women's clubs. Between 1905 and 1916, many women participated in the movement for the creation of national parks. These women were mainly conservationists. Although they were active in public life, they retained conservative tendencies which were contradictory to the progressive era that followed.

Both the environmental and the women's movements originated during 1960s and 1970s. Hence, the exploration of connections and the implication of identities between nature and women and between ecology and feminism as inevitable. While formulating their own theories, feminists grapple with women's inclusion in the category of nature in traditional philosophical literature and the reshaping of the same theme in modern philosophical literature. One strand of ecofeminism endorses the association between women and nature without critically examining how the association is produced by exclusion. Another strand of equality feminism endorses women's transition from the sphere of nature to that of culture/reason without analysing the problematic concept of 'reason' which has contributed to such exclusions. It is extremely challenging to avoid being trapped with either pole.

Some feminists, such as Mary Daly, Sherry Ortner and Susan Griffin, operate on the assumption that all recent cultures are fundamentally patriarchal and denigrate women by identifying them with nature. Others, like Merlin Stone, Adrienne Rich etc., argue that today women can recapture the early age of nature worship (represented by Mother Goddesses and egalitarian practices) through ceremonies, rituals, poetry and a new language thereby reinstating the ancient ideal.

It is important to generate symbolic responses that can lead the way to a new ethic and new behaviour patterns towards the earth as well as women. However, Merchant feels that symbolism is double-edged. The identification of women and nature, implicit in phrases like "man's war on Mother Nature", are suggestive of sexual assaults. She firmly believes that without a simultaneous revolution in social, sexual and economic structures, the symbolic revolution cannot succeed.

The psychological approach to the woman-nature question adopted by Nancy Chodorow, Dorothy Dinnerstein and Evelyn Fox Keller also locates the origins of female subordination in biological functionalism. The sexual division of labour results in a dual psychology rooting the female psyche in fusion, empathy, identification and wholeness and the male psyche in separation, distinction, division, antithesis and dualism (mind/body, people/nature, subject/object). This results in the domination of women and nature as the 'other'.

Merchant is critical of feminists like Shulamith Firestone who want to achieve women's liberation through reproductive technology. Many feminists negotiating the postmodern perspective believe in social and technological constructivism. They neglect problematizing ethical and political aspects, such as the control over life and the diversity principle essential to sustaining life on earth. Merchant cautions them against favouring the liberatory and progressive aspects of these technologies.

Struggles against Industrial Pollution

The link between the home and the environment had been established as early as the 19th century by the ecologist Ellen Swallow. After graduating from Massachusetts Institute of Technology, Swallow developed a science of environmental quality which she called "Oekology". It was concerned with industrial health, water and air quality, transportation and nutrition. Once known as home ecology, it was later diluted into home economics.

Seventy years later, Rachel Carson made caring about the earth a public issue. Her book, *Silent Spring* (1962), focused attention on the negative impact of chemical insecticides accumulating in the soil and tissues of living organisms. The chemicals were distributed as residue through surface and underground water. They progressively concentrated in the food chain thereby contaminating and poisoning all life. Carson asserted that "we stand now where two roads diverge". The less travelled fork – the road of biological control by natural enemies and of alternative control by technologies – "offers our last, our only chance to reach a destination that assures the preservation of our earth".

The first conference to establish a connection between nature and women with a programme of environmental action for the 1980s was held at the University of Massachusetts, Amherst. Several conferences followed focusing on different aspects of environmental problems. The summer institute took place on "Ecology and Feminism". Students organized classes on "Women and Nature". According to Merchant, resistance to a feminist-environmental coalition came from both movements. Environmentalists, mainly men, reacted negatively to the intrusion of feminist issues which seemed to complicate an already difficult struggle. Feminists expressed dismay at the untimely desertion of many women who joined the 'anti-nuke' movement and the environmental movement at a time when issues like ratification of the Equal Rights Amendment, anti-abortion legislation and childcare costs were looming on the horizon.

Struggle against Nuclear Technology

The resistance movements against the backdrop of the consequences of industrial activity was the beginning of the new environmental movements. In the 1970s and 80s, women's participation in the environmental movement was most pronounced on nuclear issues. The National Organization of Women adopted a resolution in October 1980 "against the use of nuclear power, and in favour of safer energy methods, including solar power, wind power, organic conversion, hydroelectric power and coal mined safely and burned safely". The movement also found a visionary prophet in Dr Helen Caldicott (a paediatrician) whose book *Nuclear Madness* succinctly describes the aftermath of a potential nuclear war.

In 1980, the Pine Ridge Hospital on the Indian Reservation (near Edgemont, South Dakota) witnessed high rates of birth defects and cancerous deaths. An investigation of the Cheyenne river revealed alarmingly high levels of radiation from uranium tailings. Women of All Red Nations (WARN) was organized to discuss the erosion of the family, forced sterilization and shrinkage of reservation lands. In the summer of 1980, the Survival Gathering of Native American People issued a "Declaration of Dependence on the Land".

Struggle against Chemical Wastes

The same period witnessed a struggle against hazardous waste dumps. The dumps were linked to elevated rates of miscarriages, birth defects and still births in surrounding neighbourhoods. Women researchers conducted studies of these problems.

The anti-toxic movement has acquired a large following among minorities. A total of 58 per cent Blacks and 53 per cent Hispanics live in communities affected by indiscriminate dumping of hazardous wastes. The movement concentrated on resisting the creation of landfills, toxic waste sites and incinerators in inner cities, poor rural areas and on American Indian Reservations. Environmental equity was considered an ecofeminist issue.

Through these struggles people became increasingly aware that the ultimate solution to such problems was reducing production (and not Superfund clean-up money). The following checklist for industries was also created: (1) reduce production of unnecessary (especially toxin-producing) commodities and focus on the fulfilment of basic needs, (2) reduce and eliminate waste output, (3) recycle, reuse and exchange waste and energy that cannot be eliminated, and (4) treat and destroy non-recyclable waste at the site of production instead of releasing it into the environment.

Struggle against Pesticides and Herbicides

Pesticides are routinely sprayed over national forests and dusted on bean and tomato fields using aeroplanes. In many places, children attending summer school were exposed to the spray and fell ill. The chemicals were suspected of causing health problems. Some of them were carcinogenic and detrimental to the nervous system. Another significant problem was the birth defects caused due to the spraying of parks with as many as 13 different pesticides.

Focus on Appropriate Technology

Many women responded by supporting the Appropriate Technology Movement which revered and disseminated hands-on skills (for personal survival and control over one's own life). Renewable energy sources, energy and food self-reliance and low-environmental technologies were its hallmark. The women involved in the movement built bridges, solar collectors, green houses and did home repairs themselves without resorting to high-cost contractors. Women taught carpentry and plumbing to other women. Many of these issues were later incorporated in the agenda for Earth Summit.[6]

A Glimpse into the Environmental Movements in Europe

By embracing common issues, environmental movements in Europe have acquired enough political strength to fight elections. Besides the well-known Green Party in Germany, they have had a political impact in other states as well. Women have been active in green parties all over Europe (particularly in the anti-nuclear movement). The most famous contemporary struggle is that of Greenham Common (UK) (see Annexure 2) in which women camped outside a missile site for a year to resist the location of missiles. The dumping of nuclear waste produced by nuclear energy plants has always been opposed in Germany and France.

A partially effectual challenge to consumerism in the West was posed by the Women's Environmental Network (WEN) in Britain. Together with War on Want, these activists organized a workshop on "Women, Environment and Development" (WED) in London. Their recommendations included an appeal to non-governmental organizations, bilateral and multilateral agencies to reconceptualize the export-led growth policy. They also pressed the governments to introduce more environment-friendly production processes.

Women, Environment and Development Discourse (WED)

Braidotti et al. (1994) map the concerns of women and environmental (in the West) in their book *Women: The Environment and Sustainable Development*. They note that in the 1970s the development discourse (led by the UN and other international organizations) was influenced by a particular stream of feminism. This school was concerned with women's relationship with the environment in the countries of the South. The oil crisis (1973) jolted people into the realization that natural resources were finite.

Development planners working on energy needs realized that a majority of people in the South depended on wood fuel for energy. A two-fold strategy was adopted to grapple with the diminishing availability of wood energy: (a) reduction of wood fuel consumption by introducing wood saving stoves, and (b) large-scale reforestation to increase wood supply. This was the beginning of the Women, Environment and Development (WED) discourse from within development-related disciplines like forestry and agriculture. It became increasingly apparent that women spent a great deal of time and energy in trying to obtain fuel, water and fodder for household use. The Chipko Movement encouraged forestry experts in donor agencies to acknowledge women as worthy of inclusion (because it was in their own interest to protect the environment). During the 1972 UN conference, the problems of the South, such as environmental degradation, a growing scarcity of natural resources and a burgeoning population, were planted firmly on the UN agenda.

The United Nations Environmental Programme (UNEP) was established after the 1972 Stockholm conference. It soon became a catalyst-cum-coordinator on environmental issues within the UN. Around the same time, the Environmental Liaison Centre International (ELCI) was set up to integrate NGO input into the UNEP. Both agencies became active in the field of WED around the mid-1980s. The ELCI organized a workshop on "Women and Environmental Crisis" at the Nairobi conference in 1985. The case studies presented portrayed women as environmental managers crucial to promoting sustainable development. Since then, the theme of women and environment has gained momentum. Drinking water and sanitation programmes highlighted women's role in this area. Consequently, the UN International Scientific Training and Research for the Advancement of Women (INSTRAW) produced training manuals for women in view of their primary role in the supply of drinking water and responsibility for their families, sanitation standards.

The Brundtland Report "Our Common Future"[7] (1987) highlighted the importance of environmental issues in the development process and proposed long-term strategies for sustainable development. Many development agencies and NGOs focused on women's involvement in such strategies and programmes. They sent female experts from the North to implement programmes and to promote the training of women extension workers in the South. WED slowly became a professional field for women development experts.

In 1987, a group of women experts from different organizations met to discuss how women's concerns could be brought into the World Conservation Strategy. In 1989, the United Nations Fund for Population Action (UNFPA) published a report called "Investing in Women: The Focus of the Nineties". The report, prepared by its Director Nafis Sadik, illustrated the link between population and environment. Consequently, many environmental groups in the North began to identify population growth as the root cause of global environmental degradation. This remains a matter of fierce disagreement between women's organizations from the South and environmentalists from the North.

Integration of Women's Concerns in the UNCED Agenda 21

The process of integrating women's concerns into mainstream environmental debate culminated with the creation of a separate platform for women – the Planeta Femea – within the NGO Global Forum (held parallel to the UNCED at Rio de Janeiro in 1992). After much controversy, the Women's Action Agenda was endorsed in the UN report. It emphasized that the successful implementation of the UNCED Agenda 21 (and all other previously published UN documents) depended on the active involvement of women in economic and political decision-making. Women's empowerment was identified as a prerequisite to sustainable livelihood for all people.

Earth Summit (United Nations' Conference of Environment and Development)

Women's policy recommendations at the Earth Summit were presented in two consecutive conferences held at Miami, Florida in November 1991. Through an exercise in cooperation and consensus, the conference formulated the Women's Action Agenda 21 to be brought to the Earth Summit. In Rio, after examining and debating the themes of the Miami Women's Action Agenda 21, the women's tent adopted the "Global Women's Treaty for NGOs seeking a Just and Healthy Planet" which was incorporated into the NGO Global Forum's final treaty.

A session at the Planeta Femea was devoted to the "Code of Environmental Ethics and Accountability" of the Miami Women's Action Agenda. Merchant avers that this Code can be used to build an alternative worldview which can be called a partnership ethic of earth care. The salient points of Women's Action Agenda 21 and its Code of Environmental Ethics and Accountability are listed below:

- Fulfilment of basic needs takes precedence over profit.
- Depletion and pollution are part of individual and corporate accounts and should be paid by the producer and polluter.
- Resources should be replenished, environments restored, and biodiversity maintained by all industries and businesses, especially transitional corporations.
- Air, water and soil should be left clean and healthy.
- Corporations, institutions, organizations, states and nations are accountable to the public.
- Environmental audit and impact assessment should be made for all proposals before funding.
- Voluntary birth control should be managed by women for women. Contraception should be safe and legal.
- Reproductive health care and family planning should be available to all women.
- Education, sex education, job education and old age security should be available to all women.
- Men should participate in childcare.

Women's environmental groups at the NGO Global Forum also succeeded in substituting the term 'sustainable livelihood' with the official slogan of 'sustainable development'. Sustainable development encodes economic growth with proper maintenance of resources. However, it does not pay attention to inequalities created in the process. Merchant emphasizes the distributive aspects of development. She defines sustainable livelihood as a people-oriented approach which focuses on the fulfilment of basic needs – health, employment, old age security, elimination of poverty and women's control over their own bodies, methods of contraception and resources. The nomenclature shows how women's perspective differs from that of men's.

Indian Environmental Movements: Forgotten Ecofeminist Component

Environmental movements in India have mainly focused on displacement due to large industrial projects and the depletion and unequal distribution of land, water and forests which deprives local communities. The origin of Indian environmental movement is traced back to the Chipko (hug) movement in Garhwal (Himalayas) in 1980s (see Annexure 3). It was an indigenous movement to protect trees which formed a principal resource for women living in the foothills of the Himalayas. The movement gave impetus to forest-based movements in India. These movements had a critical impact on the proposed Forest Bill of 1982 (this bill was withdrawn later). The greatest achievement of these movements was a paradigmic shift in the discourse on the commodification of natural resources.

After Chipko, and the Appico (hug in Kannada) movement in Karnataka, women participated in many other environmental movements. Their contribution is often ignored because they were apparently not on the forefront. Women's lives are directly affected by population displacement and degradation of the environment. They invariably constitute large contingents in protests on environmental issues.

These environmental movements differ from movements in the North. Unlike the North, our debate on the ecological crisis did not give impetus to philosophical questioning, such as the relationship of humans with nature. It focused either on finding the best path towards development or on alternative visions of society. Activities under the rubric of 'Women and Development' are strongly resented. The main concern is the sustainable use of natural resources while ensuring livelihood for people dependent on receding supplies of the same. An ecological crisis here is felt more harshly and directly. Such situations often result in social conflicts because different class groups exercise competing claims on a dwindling resource base. These conflicts are played out at different levels and with varying intensities.

Wide Spectrum of Issues and Movements

Janki Andharia and Chandan Sengupta list environmental movements into eight types according to the resource base they are responding to (Andaria and Sengupta, 1998: 429–31). Their distinction has been maintained and elaborated upon (using illustrative examples from activism in India) in the following list:

1. *Forest and Land-based Issue:* Chipko, Appiko and tribal movements across the country demand access to forest resources along with a ban on the commercial felling of trees. They seek to preserve land in the hills through forestation. Dryland areas are increasingly demanding watershed development in places where the sourcing of underground water for commercial crops has emptied local wells and made drinking water a scarce resource.

2. *Marine Resources, Fisheries and Aquaculture:* National Fishermen's Forum (working for traditional fisherfolk in Kerala) has been demanding a ban on trawling near the sea shore and during the prawn regeneration season. Women who derive their livelihood from selling fish have to travel deeper into the interior to locate customers (because of large companies dominating city markets). Sweet water prawn culture in Chilka Lake and in the paddy fields of Andhra Pradesh has engendered agitation against commercialization as it degrades land within a few years.

3. *Industrial Pollution:* Common demands to the industrial sector include stricter pollution controls and compensating accident victims. Zahiri Gas Kand Morcha (Forum of the People Affected by the Poisonous Gases) in Bhopal has been actively seeking compensation for the victims of the Union Carbide tragedy where more than 3,000 people died. Many women had miscarriages or gave birth to children with impaired health.

4. *Development Projects:* Dams, irrigation and power projects, mining, industrial plants, railway, airport and military bases etc. cause displacement. The Silent Valley Movement

started by the Kerala Shastra Sahitya Parishad (KSSP) shed light on the destruction caused by development projects. It became a well-known issue by the time the Narmada Bachao Andolan was launched. Tehri Band Virodhi Samiti in Garhwal, Koshi Gandhak, Bedthi and Ichampalli in the east, the Tungbhadra, Malprabha and Ghatprabha schemes in the south and the Koyana project in Maharashtra have also raised forceful voices.

Examples of resistance against displacement include the Jan Andolan against the Dabhol Power Plant (being set up by Enron – an American power installation company) in Maharashtra and the anti-nuclear energy agitation against the Kaiga project in Karnataka. They highlight the issue of development benefiting the rich at the cost of the poor. Other examples include anti-mine projects in Dehradun Valley and Balco (Orissa), protests against missile test ranges in Balliapal (Orissa) and Nethral (Bihar), the Konkan Railway Realingnment Action Committee and the Citizen's Group against Dupont Nylon 6.6 in Goa.

5. *Wild Life Sanctuaries and National Parks:* Ekjoot in the Bhimashankar region of Maharashtra and Shramik Mukti Andolan in Sanjay Gandhi National Park (Mumbai) prove that maintaining sanctuaries need not include displacing tribal people. Tribals have preserved forests for centuries by adjusting their lifestyle to the forest's rhythms. They cannot be rehabilitated without harming their cultural practices and community life. Their traditional livelihood cannot be supported in any other environment.

6. *Tourism:* Himalaya Bachao Andolan in the north and Bailancho Saad in Goa have been canvassing against tourism as an industry as it introduces alien consumerist lifestyles and devours scarce natural resources. This deprives local people of their livelihood. Bailancho Saad accuses the state of trying to attract tourists by creating an image of "free women" thereby taking advantage of the historic influence of Portuguese culture on Indian women.

7. *Appropriate Technology:* Advocacy groups respond to issues at the macro level (by pressurizing the government to adopt policies which reduce environmental damage and to develop appropriate technology etc.). Advocacy hopes to change social relations by creating alternatives to current development processes. The Centre for Science and Environment in Delhi, Bombay Natural History Society and the Save Pune Citizens' Committee are organizations providing advocacy. Projects in the appropriate technology field include Laurie Baker's housing experiments, Ralegaon Sidhhi's ecosystem-based development and CASAD's (Centre for Applied Systems Analysis and Development) experiments with materials which maximize the biomass content of the construction. Groups concerned about losing land to the heavy use of chemical fertilizers and to excessive irrigation for commercial cropping are advocating organic farming.

Andaria and Sengupta mention that many economic projects under the rubric of 'development' affect millions, especially the rural and marginal populations. However, their benefits are reaped by a few rich, urban Indians. Companies and governments from the North externalize the environmental costs of their economic activity by exporting polluting industries to the South. The elite in the South export these activities to their own fringes, i.e., rural areas where people are not as organized. Dasmann calls the elite in the North and South 'biosphere people', i.e., those who enjoy the produce of the entire biosphere (Dasmann, 1988). In India, masses of people have to sacrifice their traditional means of livelihood for 'progress' and 'development' which remains elusive to them.

Despite women's participation in movements, their demands are not always articulated separately. Women's responses are not studied in a sensitive manner and their ideas are often ignored or excluded. This gap gives the impression that simple living and stability of land are a romantic vision nurtured by middle class activists, which is imposed on seemingly awakened rural people, especially women. Many studies have documented the impact of environmental degradation on women's health, nutrition and cultural networks. Taking their solutions seriously can offer a refreshing perspective on the problems of development.

Ecological Crisis: Contradictions Caused by the Nature of Resource Use

Madhav Gadgil and Ramchandra Guha classify Indians on the basis of their resource use. This framework is used to analyse conflicts caused by practices competing for priority within the same framework. Understanding how people from different categories relate to an ecological crisis helps to explain the responses generated by environmental movements (Gadgil and Guha, 1998: 448–51). In their classification, India's population is divided into three categories: 'omnivores', 'ecosystem people' and 'ecological refugees'. These three classes are distinguished on the basis of the size of their respective resource catchments or by their relative ability to transform nature into artifacts. People can be classified on the basis of their power to influence state policy or by the degree of control they exercise over their own lives. *Omnivorous people* are defined as those who devour everything produced anywhere on earth. *Ecosystem people* have limited resource catchment. *Ecological refugees* are those who have been displaced by dams or missile ranges. They are lost and marginalized in urban fringes.

Gadgil and Guha (ibid.) assert that the environmental movements in India (formed by members of one or more categories of the aforementioned triad) fall into seven major strands. Two of these focus exclusively on nature conservation movements, one on aesthetic/recreational/scientific grounds and the rest on the basis of cultural or religious traditions. The first strand, called the wildlife conservation movement, attracts urban omnivores. The second strand is represented by the Bishnoi peasants in Rajasthan who protect Khejadi trees and blackbuck, chinkara, nilgai etc., around the village.

The third, and the most dominant, strand stresses on equity issues. These movements arise out of conflicts between omnivores (who gain disproportionately from economic development) and ecosystem people (whose livelihoods are seriously undermined through a combination of biased resource fluxes and growing degradation of the environment). The authors call these movements 'environmentalism of the poor'. The remaining four strands identified within them are Gandhians, Marxists, concerned scientists and spontaneous efforts of local people. The efforts of local people include attempts to protect and use local woods or ponds sustainably or to pursue environment-friendly agricultural practices. The Gandhians are motivated by moral imperatives such as checking overuse and justice for the poor. Marxists emphasize the dismantling of the social order through struggle. The scientists advocate reconstruction and deploying appropriate technologies. People's initiatives are hard to come by, but they do encourage sustainable practices.

Nature of Struggles

Environmental struggles can be distinguished into three different sets of initiatives which are sometimes interrelated. Firstly, some initiatives try to prevent ecologically destructive economic practices through a process of organization and struggle. These have met with varying degrees of success. Secondly, some initiatives promote the environmental message through the skilful use of the media and by using informal means, such as walking tours and eco-development camps etc. Thirdly, social action groups have taken up programmes of environmental rehabilitation (e.g., reforestation and soil conservation), restoring degraded land and village ecosystems.

Gadgil and Guha (ibid.) aver that these efforts are adopted by groups outside the sphere of formal politics. They further emphasize that the entire ideological spectrum of politics in India has turned a blind eye to the continuing impoverishment of India's natural resources. Environmental activists have questioned the orientation of economic planning, its in-built biases towards the commercial/industrial sector and its neglect of economic considerations. Although perspectives within the movement are varied, this public debate on development options constitutes the ideological expression of the environmental movement.

The authors categorize groups according to their visions for future society and strategies for struggle towards this end. Thus, they are classified under broad titles such as Green Development, Crusading Gandhians, Ecological Marxists, Appropriate Technologists, Wilderness Enthusiasts and Scientific Conservationists. Labelling initiatives and organizations is mechanistic but it shows that they differ in orientation. A few examples are cited here to illustrate these differences.

The objectives of Jan Vikas Andolan, started concurrently with the rally against the Sardar Sarovar dam at Harsul in 1989, are four-fold:

- to coordinate collective action against environmentally destructive policies and practices;
- to provide national solidarity to these struggles;
- to mobilize wider public opinion on the need for a new development path; and
- to work towards an ecologically sustainable and socially just alternate vision for India's future.

The Jan Vikas Andolan and the National Alliance of People's Movements have been networking with local struggles and are taking a political stand against the present development model. According to them, the current model is resource illiterate and directly responsible for impoverishing the resource base of rural people. There have been very few attempts at working out plausible alternatives.

Crusading Gandhians have carried their message of moral regeneration across the globe. They attack the Indian intelligentsia's obsession with modernist philosophies, particularly those upholding rationalism and economic growth.

Ecological Marxists are identified with the People's Science Movements (PSMs) in various states. PSMs' initial concern of taking 'science to the people' has been extended to include environmental protection. PSMs are hostile towards tradition and nurture the corresponding faith in modernity and modern science. They favour confrontational movements.

Appropriate Technologists strive for a working synthesis of agriculture and industry, big and small units. They adhere to technological traditions and have pioneered the generation and diffusion of resource conservation. Appropriate Technologists favour labour intensive and socially liberating technologies. They emphasize practically demonstrating socio-technical alternatives to centralized and environmentally degrading technologies.

Wilderness Enthusiasts, according to Gadgil and Guha (ibid.), enjoy the support of omnivores. Indian naturalists have extensively documented the decline of natural forests and their plant and animal species. They have urged the government to take remedial action. Recently, they raised the issue of biological diversity by deploying a moral argument favouring 'species equality'. This was done to create a more extensive system of parks and sanctuaries and to advocate a total ban on human activity in protected areas.

Scientific Conservationists are an influential strand within the state. They focus on efficiency and considerably influence government policy. For instance, the pioneering bureaucrat B.B. Vora has documented the extent of land degradation, water erosion etc. Many laws were passed due to the influence of Scientific Conservationists and Wilderness Enthusiasts. They are criticized by Gandhians and Marxists of being elitist as they fail to recognize the social consequence of environmental use and abuse.

Ecofeminism: A Blindspot in the Documentation

Interestingly, Gadgil and Guha (ibid.) do not acknowledge ecofeminism. The weakness of the ecofeminist strand in India is compensated by the forceful voice of Vandana Shiva. She is active both at the grassroots and at the international level. In addition to her prolific writings (including the book *Staying Alive* published in 1988), she is associated with the Rayat Sangh – a farmers' movement in Karnataka. The movement is directed against Kargil and Monsanto who test genetically engineered seeds in the region. Shiva has been instrumental in setting up seed preservation plots (using *in situ* methods) where farmers can buy traditional varieties of seeds. Her primary concern is the preservation of biodiversity, particularly staple food grains. A few years ago in Delhi she inspired many women's organizations to come together to protest the signing of WTO agreements regarding the trade of food grains. Shiva warns that the WTO can potentially destroy our independence in producing and supplying food. It is a threat to food security within India. Her scathing attack on seed companies, which distribute 'terminator seeds' thereby increasing reliance on pesticides and fertilizers, has aroused the interest of urban women as they manage the food consumption of the family.

No environmental group has tried to consider the place of women within an ecological paradigm on the basis of their productive and reproductive activities (pertaining to food and health). Women are not mentioned when a concern for equity is expressed. Traditional injustice prevails over women even through government policies. The denial of land inheritance rights is compounded when common property resources are threatened by degradation and privatization.

The denial of access to critical natural resources has a disastrous impact on women. Despite local protests, their voices have not inspired environmental groups to demand land and water rights for them. Women figure in a very limited manner within the Joint Forest Management (JFM) system or in Watershed Committees. Still these groups rely on the family as a unit for redistribution programmes. Ecofeminism will not be treated like a viable and visible strand by other movements unless a movement demanding separate rights and accesses for women is built into them.

Women's relationship to resource use is more specific within the categories of omnivores, ecosystem people and ecological refugees. There is ample documentation to prove that in any category, women's response to an ecological crisis may not be the same as that of men. Within the Chipko movement itself, men had demanded felling rights for local men (rather than outside contractors) and had expressed interest in mining limestone in the same region. Women wanted the trees for use as they were. Hopefully, further documentation will prove that women constitute a subcategory within each broad category and hence demand special strategy formulation in popular resistance as well as in government policy.

Annexure 1
The Emergence of Environmentalism
(Huges, 2005: 868)

- Environmentalism as a popular social movement emerged during the 1960s. It is regarded as a response to the publication of Rachel Carson's (1962) book *Silent Spring*, which warned of the dangers of pesticides such as DDT.
- It was expressed in the first Earth Day held on 22 April, 1970. Initially, the Earth Day was primarily an event in the US. It later received international observance.
- The United Nations Conference on the Human Environment (Stockholm, Sweden) in 1972 was the first major international conference devoted to environmental issues. The only heads of state present were Sweden's Prime Minister Olaf Palme and Indira Gandhi (India) who was the spokesperson for developing countries.
- Would measures for conservation of resources and reduction of pollution limit the development of poorer countries while leaving the richer countries in relative affluence? Gandhi maintained that poverty and need are the greatest polluters and that the basic conflict is not between conservation and development but between the environment and the reckless exploitation of humankind and nature in the name of economic expansion. He insisted that discussion of environmental problems be linked to issues of human needs.
- The Stockholm Conference led to the creation of the United Nations Environment Programme (UNEP), headquartered in Nairobi (Kenya), for coordinating the efforts of United Nations on the environment.
- In addition to seeking reform in governmental action and international agreements (which seemed to be slow and inadequate) some environmental movements also organized protests against hazards like toxic waste dumps and polluting industries located in neighbourhoods inhabited by poor and racial minorities or clear-cutting of ancient forests.
- In 1993, Canadian citizens blocked the roads being used by the logging trucks of the company MacMillan Bloedel which transported trees cut from the forests of Vancouver Island.
- The Chipko Movement began near Himalayan villages (northern India) in March 1973. The villagers needed trees for fuel, fodder,

small timber and protection against flooding. Peasants prevented loggers from cutting down trees by threatening to hug (*chipko*) the trees and place their bodies in the path of the axes. Many demonstrators were women who gathered wood. Similar protests occurred elsewhere in India and met with some success.

- In Malaysia, the Penan people of Sarawak carried out a series of direct actions including roadblocks to protest the destruction of their rainforest by commercial loggers.
- People who have demonstrated concern for the relationship of humans with nature have often been persecuted:
  - (a) Wangari Maathai was beaten and imprisoned for starting the Green Belt Movement in Kenya. This movement enabled women and children to plant trees and nurture them.
  - (b) Judi Bari, the leader of the 1990 "Redwood Summer Protests" against logging of giant redwood trees in California, was maimed by a bomb placed under her car seat.
  - (c) Chico Mendes organized *seringueiros* (rubber tappers in Brazilian rainforests) to defend the forest and their livelihood against illegal clearing. He was murdered in 1988 by agents of rich land owners whose financial interests were threatened.
  - (d) Ken Saro-Wiwa – poet and playwright – was executed by the Nigerian dictatorship in 1995 for organizing a movement against oil drilling which produced air and water pollution in the lands of his Ogoni tribe without compensation.
- Two well-known worldwide organizations related to conservation efforts are World Wide Fund for Nature (WWF) and Greenpeace. WWF was launched in 1961 and funds projects in five categories: individual species, wilderness areas, support for existing organisations, conservation education and miscellaneous conservation matters. Greenpeace predominantly works for sea mammals and the elimination of toxic pollution.

Annexure 2
Greenham Common: Women at the Wire

The Beginning

- The Newbury Corporation purchased Greenham Common for the enjoyment of local people. However, after World War II, it became an air force base for the Royal Air Force. Even after falling into disuse, it was not returned to the Council for public use. It was sold to the Ministry of Defence which leased it to the American Air Force.
- The protest was against American Cruise missiles coming to Britain in accordance with a NATO decision taken in December 1979.
- The Women's Peace Camp at Greenham Common was not planned. People gathered because of the 'Women for Life on Earth' march from Cardiff to the United States Air Force (USAF) base at Greenham on 27th August, 1981. Frustration at the media's disinterest, the determination of the marchers and the enthusiasm of supporters combined to create a continuing outdoor settlement which survived three winters, evictions and attacks.
- The connection between the nuclear arms race and the structure of the society are described in terms of 'patriarchy' by Jayne Burton who maintains that within a male-dominated society, male-dominated institutions and stereotyped male values determine resources use.
- Women identify with Greenham because they witnessed women taking initiative that built on women's strength rather than denying it.

Features of the Women's Peace Camp at Greenham Common

— It began as a predominantly white, middle-class movement which broke the barriers of race, class and sexuality as it grew.

— From the very beginning of the camp, the principle of working without hierarchies was established. Women adopted a non-violent strategy for the peace movement. The camp adopted routines which ensured that women could speak and act for themselves.

— Living conditions were primitive – without electricity, telephone, running water etc. Outside the main gate of the camp, there was a large communal fire and a common tent. There were caravans, tipis and tents. Banners were put up near the highway and the entrance of the main gate. In spite of the conditions, women from all over UK and abroad came to spend time at the camp to be part of the resistance to nuclear weapons.

— Support and donations came in the form of visits, money, food, clothing, stationery, tents etc.

— A number of initiatives testing the legality of nuclear weapons were made by women in court. Many of them faced court cases, fines and sometimes imprisonment for their actions.

Some of the Events Organized at the Camp

- In December 1981, a part of the camp was to be bulldozed for laying sewage pipes for the air base. Women sat down in front of the digger and stopped the work. This incident forced the women to think out their attitudes towards confrontation.
- On New Year's Eve 1981, a few women walked around the base and hung painted and written messages along the fence. Sometimes a message was left at a single house or a place of work inside the base.
- The preparations for the Cruise missiles coming to Greenham were hindered and disrupted. March 21st was celebrated as the Festival of Life. The first blockade was organized on the next day. This was the first time women at Greenham tested non-violence on a large scale.
- On 27th August, 1982 some women were charged for occupying the sentry box at the main gate.
- On December 12th women 'embraced the base' (i.e., encircled the base by linking hands). The event was publicized using pamphlets, advertisements in newspapers etc. Nearly 30,000 women came even though only 16,000 were needed to encircle the base.
- A dance on the silos took place on 1st January, 1983. A total of 44 women were tried for breaching the peace. All of them were found guilty and imprisoned for 14 days.

- In May 1983, an attempt was made to evict the Women's Peace Camp. Women retaliated by blockading the roadway leading to the main gate. They lined across it with their arms linked.
- International Women's Day for Disarmament was celebrated on 24th May, 1983. The coordinating office in London informed the Greenham Peace Camp that "over a million women took part in thousands of actions all over the world".
- On 29th October the fence around the base was cut using blotcutters.
- The Ministry of Defence, Ministry of Transport and the Newbury District Council combined forces to commence evictions (starting at north side of base) in preparation of Cruise convoy deployment on 5th March, 1984.
- The conduct and integrity of the protest mounted by the Women's Peace Camp was instrumental in removing the Cruise missiles from Greenham Common. The camp remained in place till 2000 when protestors won the right to house a memorial on the site.

Diversity in Point of View

— From the beginning of the camp there were terrible scenes over the women-only/mixed protest issue. Initially men did an ample share of the menial tasks and stayed in the background while women gave talks and handled the media. One of the fears was that men readily respond with violence in a dangerous situation because of their conditioning and protective feelings towards women. This would take away women's responsibility for their own actions. It was also felt that the police may respond in a more heavy-handed way towards men.

— National and international media descended in force to isolate spokespeople who could provide one-liners to define the camp and the protesting women. This caused disagreements between women.

Annexure 3
Women's Role in Chipko Movement in Uttar Pradesh, India (Jain, 1984: 1788–1974)

Introduction

The Chipko movement is an ecological movement concerned with the preservation of forests and maintenance of the traditional eco-balance in the sub-Himalayan region. It strives for the traditional status quo between people and the environment. The proponents of this movement have demonstrated the negative effects of the previous and current forest policy adopted by the Indian government on the ecological balance of the area. This policy uproots indigenous people who are dependent on forests.

The collective mobilization of women (for preserving forests) has caused a conflict about their status in society. Men's opposition to women's role in the movement stems from the women's demand for sharing in the decision-making process.

- Why do women support the Chipko movement?

Women (being the sole in-charge of cultivation, livestock and children) lost everything to recurring floods and landslides. They realized the cause of their problems. A sheer need for survival made them support a movement which sought to preserve the ecological balance in the area.

- Why didn't men make the connection?

This involves two reasons: one is the way in which subsistence economy is organized in these areas and the other is men's preference for alternative development taking place in the country.

A brief account of the social setting of the region:

— Local economy: The Chipko movement was initiated by a group of Sarvodaya workers in the Garhwal region (UP).

— Social status of women and men:

- Most families own a small patch of land for cultivation (covering an area of less than an acre) and hence there are very few agricultural labourers. Farming is mainly dependent on monsoons rather than irrigation.

- More than 95% of working women are involved in cultivation. They do not work in mining, quarrying, transport etc., while only 72% of the men are cultivators.
- Single member female households outnumber single member male households. Male migration (from hilly areas to the armed services and other jobs in the plains) is fairly common leaving women to look after land, livestock and families. A large number of households are run by females in the absence of male heads. In some cases women head the household even in the presence of male spouses.

— Role of forest in the life of people:

- Settled agriculture is coupled with foraging of minor forest produce. The forest is used for producing agricultural tools, dwellings, cooking, fuel, grazing cattle, food etc.
- An average family of five members can survive on farm produce for 3 to 6 months. For the rest of the year, the villagers sustain themselves using other sources (primarily the forest).
- People had free access to the forest till 1821. Since then, a gradual process of usurping control has been initiated by the government.

— Sex-role differentiation and relation between women and men:

- They are linked to the pattern of cultivation and exploitation of forest wealth. Women's position in society is governed by the norms of the patriarchal system of social organization.
- Men own the land since property is transmitted patrilineally. Control over the allocation of the patrilineal group's resources is solely in male hands. Senior men govern land, livestock, political and ritual offices, control over marriage of sisters and daughters. Important rituals are organized by men. Women say that their work prevents them from being a part of the political and ritual life of the community.
- Men prepare the land for cultivation as there are taboos associated with women operating the plough. Preparation of the soil does not

give men the right to distribute food in or outside the family; it is generally in the hands of women.

- Men who do not migrate for work participate little in cultivation apart from preparing the land for planting. Instead, they build and repair dwellings, are involved in marketing and participate in socio-religious and political activities of the community.
- Crops are raised solely by women. They are responsible for planting, weeding, harvesting, rearing cattle and transporting crops from the fields. They cook, rear children and collect firewood, water, fuel and fodder for cattle. There is no obvious division of labour for crafts. Women exclusively control and manage food as they are the ones who cultivate crops for domestic consumption.

— Short history of Chipko movement:

- The movement started at Mandal in Chamoli district on April 24, 1973.
- It was founded by C.P. Bhatt and his co-workers (who were the natives of Chamoli district). They believed in the ideology of non-violence advocated by Mahatma Gandhi and Vinoba Bhave. They established the Dasholi Gram Swarajya Mandal (DGSM) in 1964 for starting small industries which used forest resources.
- In 1970 and 1973, hundreds of homes were swept away due to heavy floods in the Alaknanda river. DGSM workers launched a relief operation during which they realized that the floods were primarily caused by soil erosion which was linked to the clear-cutting of mountain slopes by lumber companies. Flood relief added a dimension of preserving the environment.
- Employment for the local population, reservation of their rights over minor forest produce and the conservation of forests are interrelated problems. The last problem connected the movement to a larger population and included women alongside men.
- Chipko struggle at Reni: In 1974, the forest department announced an auction of trees in the Reni forest. C.P. Bhatt reminded the villagers that the 1970 floods had been caused by felling down of trees. He suggested hugging trees as a tactic to save them. A woman called Gaura Devi organized the women of her village

against the men of the lumber company which finally retreated. The government ultimately banned tree-felling in the area.

- After the Reni success, C.P. Bhatt and his workers began addressing women and found them sensitive and responsive to ecological problems.
- Chipko demonstrations were held elsewhere in the area. For example, Kedar Singh Ravat joined hands with Sarvodaya workers and organised Chipko demonstrations in Phata. Under the guidance of Sunderlal Bahuguna, the Uttarakhand Sarvodaya Mandal in Tehri Gharwal spread the message of Chipko movement in the whole Himalayan region.

3

Environmental Degradation and Women in India

An Overview

> Encroachment into grazing lands and forests has pushed women to uneconomical use of cow dung as fuel which not only separates it from the agricultural cycle but also uses a lot of labour time. According to an estimate, in this area, almost 30 per cent of fuel needs are met by cow dung which could otherwise have been used for composting or manure, or both fuel gas and manure through biogas technologies. (Jeffery, Jeffery and Lynn, 1989)

Several documents in India explore the correlation between women's status and the impact of environmental degradation. These documents are concerned with the effects of industrial and rural activities, such as deforestation, green revolution etc. Displacement due to big development projects (dams, energy projects, missile range etc.) exerts a special impact on women. This chapter focuses on the impact of the degradation of natural resources on rural women. Livelihood security and food security are essential for rural women especially because of male migration.

To begin with, the chapter highlights the correlation between women's status, food security and environmental degradation at the macro level. Subsequently, it illustrates the positive and negative effects of the green revolution. The writings of G.S. Aurora (1991), Vandana Shiva (1989), Utsa Patnaik (1993/96), Maithreyi Krishnaraj (2007, 2008), a paper by K.S. Gopal and Shashikumar (1993), Bina Agarwal (1992), Gita Sen and Karen Grown (1985) have been used as reference.

Links between Women's Status, Health, Livelihoods and Food Security

Women's relationship to agricultural technology has been discussed since the eighties. Initially, the focus was on employment and social status as opposed to environmental degradation. It was generally concluded that new agricultural technology (termed the green revolution)[8] had an overall negative impact on women in terms of employment opportunities, wages and control over cash incomes. The exception to this rule were women in large landholding families. Technology had polarized women into 'haves' and 'have nots'. A majority of women were pushed into a more deprived status. Pointers indicated a correlation between commercial agriculture and an increase in the incidence of violence within the families of rich farmers.

Meena Shensu and Vasant Bhosale (1990) explored the negative impact of the decline in the Female Work Participation Rate (FWPR) in their booklet on violence. This booklet was based on studies in the sugar factory area of Sangli district (Maharashtra). The area was relatively affluent because of sugarcane cropping which allowed farmers to withdraw their womenfolk from cultivation. Conversely, this also increased the demand for dowry, desertion for insufficient dowry and dowry deaths. Barbara Miller's study (1985: 52–54) has documented the correlation between commercialization of agriculture and a decline in the female population (in the age group 1 to 5) in Punjab. A similar correlation is also discernible in the Kellar community (which resides in the command area of a dam in Salem, Tamil Nadu) between female infanticide and the prospect of commercial agriculture (Krishnaswamy, 1988).

Development Alternatives with Women for a New Era (DAWN) was a document published during the Nairobi conference in 1985. It made people aware of the multidimensional impact of a 'reproduction crisis' due to environmental degradation. It emphasized the following dimensions (Sen and Grown, 1985: 43–67, 71–82):

1. Women are the main producers of food crops in many countries.

2. Where women do not cultivate, they are the main food processors and cooks.

3. Women are often responsible for providing water and fuel.

4. Women have to trade off basic needs in the use of resources such as labour time, cash income and land under their control.

5. There has been a shift towards less nutritious but more labour and fuel intensive foods.

6. Women (as managers of intra-household food distribution) usually eat after men and thus consume less food.

Each of these aspects has been researched in India and can further help new researchers and policy makers in understanding the impact of environmental degradation.

After studying the literature on agricultural growth, Bina Agarwal, in her article "Women, Poverty and Agricultural Growth in India" (1985), finds that field performance of agriculture is not a decisive factor underlying trends in rural poverty. However, success in agricultural growth may have an indirect ameliorating effect. She draws attention to the association between changes in economic variables and the prevalence of factors such as infanticide and dowry payments. She further examines the implications of poverty in different geographical areas on rural women. Agarwal focuses on sharing the burden of poverty within the household by analysing the intra-household distribution of food and the relative access of household members to medical care. She infers the existence of gender inequities (due to underlying economic, social and historical differences) by tracing backwards from the incidence of malnourishment, morbidity and mortality.

Maithreyi Krishnaraj notes that poverty has increased in farming households after the implementation of liberalization, privatization and globalization (LPG). She quotes G.S. Bhalla (2006) who found out that the average monthly per capita consumption of farming households in India was Rs 503, which is Rs 9 lesser than that of rural households as a whole (according to the 59th round of NSSO). The consumption of food items (excluding cereals) like pulses, vegetables, fruits, fruits, eggs and meat was consistently lower in farming households. She observed that energy derived from food has decreased considerably among the rural and urban poor.

Green Revolution, Degradation of Lands and Effect on Food Security

G.S. Aurora (1991: 119) comments that the green revolution strategies pushed in the dryland were doomed to failure. Efforts in this direction were ceased when people realized that the green revolution technologies (despite their power to produce immediate results) created the following long-term ecologically dangerous conditions:

1. The inevitable erosion and depletion of topsoil, pollution of water (both underground and surface reservoirs) particularly due to pesticides and the oversupply of nitrogen to lakes and rivers (Jackson, 1980: 14–26).

2. An increasing dependence on commercial and industrial systems including seeds, fertilizers, implements and power. These products were formerly self-generated but become commodities bought from the market. The power to manipulate the price of the inputs rests with market leaders. When the structure is modified by a government subsidy system, the payers/consumers pay for system inefficiencies (Mooney, 1980: 43–44).

3. The modern system of agriculture uses large quantities of non-renewable energy resources. As these resources get depleted, the cost-output ratio becomes uneconomical and therefore unsustainable (Doyle, 1985: 261–70).

Green revolution technologies were severely attacked by Vandana Shiva because of their destructive developmental paradigm. According to her, environmental degradation is a break from a holistic lifestyle, i.e., from people living harmoniously with nature. She is critical of women being devalued and their relationship with nature being severed. In her book *Staying Alive* (1988), Shiva elucidates women's contributions (based on traditional methods) to the food cycle. She claims that importing Western green revolution strategies to India destroys women's relationship with food production. Women have played a major productive role in sustainable agriculture, particularly in maintaining the food cycle. They are responsible for feeding animals using trees or crop by-products, nurturing cows and other animals, composting and fertilizing fields with organic manure and managing mixed and rotation cropping. In a hectare of farmland in a hilly region, a pair of bullocks works for 1,064 hours, men for 1,212 hours and women for 3,485 hours (Shiva, 1988: 108–109). In a conversation, Krishnaraj observes that in Maharashtra, tribal women prepare 'raab' – a mixture of cow dung and crop residues which is burnt and then spread on land after ploughing.

Despite severe criticisms, many economists (including Utsa Patnaik, 1993) defended the green revolution on the grounds that it enabled us to feed a growing population and increased our autonomy in the world market. It created a bulk supply for the Public Distribution System (PDS) which is important to both urban and rural areas. Women are concerned about food security and hence (at least for their sake) the green revolution is justified. In a recent paper presented at a workshop conducted by ILO/National Commission for Women (in 1993 in New Delhi and later in *EPW*, 1996: 2429–2449), Utsa Patnaik argues in favour of the need for a PDS. She indirectly arrives at the conclusion that the green revolution technologies seem to be the solution for bulk supplies of food grains. Patnaik however does not discuss the long-term impact of such methods on the soil. These include decreased productivity and hence decreased supply.

Patnaik has studied the shift in land use patterns, particularly in Third World countries, at two points of time: (a) the colonial period in which the mother country sought agricultural raw material supply from the colonies for industrial production, and (b) the current model of repaying debts under structural adjustment programmes. Market forces separate land from the production of necessities for the poor. This reduces food grain availability and destroys sustainable agriculture. The country becomes dependent for staple food supplies on the same country which pushed market-oriented production. Such dynamics are convenient for developed countries as they are proficient in the mass production of wheat and rice.

Patnaik argues that food dependency makes developing countries vulnerable to political pressure. Countries which implement structural adjustment programmes have to compete with other Third World countries (while exporting agricultural products) in order to repay foreign loans. This creates a distress sale situation which is advantageous to First World countries but threatens food security in debtor countries. The introduction of industrial agriculture initiates a degradation process. Poverty is induced as land gets concentrated in fewer hands. People deprived of natural resources become dependent on poverty alleviation programmes which turn them into wage labourers. These labourers are supplied food grains through PDS thus establishing a vicious circle.

Patnaik notes two important features of the Indian food economy and the PDS: (a) the fragility of the country's self-sufficiency in terms of food, and (b) the implications of a high degree of regional concentration of food output. These situations make India politically vulnerable to regional secessionist movements in the short term and to outsiders in the long term.

It has to be also noted that the European Union countries subsidize their agriculture and allied activities like dairy so that they sell their agriculture products cheaper than the Indian products. The latter do not have the technology of Western Europe as bulk of our farmers are 'subsistence farmers'.

In the aforementioned DAWN document, Gita Sen and Karen Grown discuss two aspects of food production and trade. The first emphasizes the expansion of international trade while the second foregrounds national self-reliance and developing trade within a country (Sen and Grown, 1985: 46). The first approach is inherently dangerous because the international grain market is controlled by a few grain merchants. Thus purchasing food during national scarcity or emergency is almost impossible. In a country facing unbalanced payment deficits and the burden of external debts, excessive reliance on outside food renders national food availability vulnerable to external pressures. Krishnaraj retorts to the argument of mainstream economists by questioning the wisdom behind switching from food production to cash crops. While higher income can enable one to buy food either locally or from imports, is it fair to throw away nature's bounty and instal glass houses to grow flowers for Europe? A large number of landless and landed poor do not have the purchasing power to induce the market to import basic food rather than luxuries. When it comes to agricultural commodity exports, 3–6 firms control 85–90 per cent of trade. While they buy products at lower prices, we are charged a price higher than their marginal cost (Krishnaraj, 2007: 38–40).

The counter-argument to the inevitability of the green revolution strategy for self-reliance comes from a study conducted by K.S. Gopal and M. Shashikumar (1993). They present statistics to prove that the benefits of green revolution are short-lived. The growth rate of food grain production in India (computed on the basis of quintennual averages starting from 1970–75) has been declining continuously. It was 3.44 per cent per annum during 1970–75, 3.11 per cent during 1975–80, 2.84 per annum during 1980–85, and only 2.34 per cent during 1985–90. If this rate continues, the growth rate of food grain production during the period 1990–95 may turn out to be slower than the growth in the population. They conclude that the growth rate of food grain production in the post-green revolution period is lower than the pre-green revolution period. A possible reason is the shift in cropping patterns due to market forces. Food production seems to be the first casualty. For example, many rice fields are being converted to fish ponds in the delta region of Andhra Pradesh. Concomitantly, new areas for rice cultivation are not being developed as a copious water supply is unavailable and the costs of irrigation projects are skyrocketing.

Gopal and Shashikumar argue for a decentralized PDS based on farmers providing food security for each other through dry land farming. They conducted an experiment with sixty farmers for three consecutive years and tried to include a drought year in the production cycle. Their conclusion was that dry land farming provides food security if a suitable credit policy is developed and if a group of farmers join the scheme. Fallow land becomes productive thereby increasing the availability of food crops in the long run.

In many regions women have learnt to build grain banks.

Forest Policy and Impact on Women

In her book *Ecology and the Politics of Survival,* Vandana Shiva (1991) argues that tropical forest resources are undergoing the world's most severe ecological and economic crisis. The food security of the entire world depends upon linkages between the ecological stability and the economic well-being of the majority of the world's people who live in the tropics or the Third World.

Shiva explains the crisis of tropical forestry in the light of the conflicting demands for biomass by the three fundamental economies associated with forests:

1. Nature's economy of essential ecological processes.
2. The survival economy of basic needs satisfaction for people.
3. The market economy of industrial-commercial demands.

Nature's economy of essential ecological processes generates a demand on the products of the forest in terms of the maintenance of soil systems and the hydrological balance of the forest ecosystems. The survival economy of basic needs satisfaction reflects the requirement of forest biomass (wood fuel, fodder, fruits, nuts, green manure, small timber etc.) by people living in and around forests. The market economy of industrial-commercial demand converts forest biomass into items of the formal market economy, such as pulpwood, plywood, furniture, housing timber, wood fuel etc., for the urban market.

Nature's economy and survival economy have always overlapped in India. They functioned simultaneously without conflict as the survival needs of local people were satisfied through conservation-oriented utilization.

The colonial rule introduced dramatic breaks in usage. The former perception of the multiple functions of a forest ecosystem was replaced with a one-dimensional scientific forestry whose only objective was maximizing production of commercially valuable timber. Market logging eliminated the utilization of forest resources for ecological or economic objectives. Post-colonial forest policy further aggravated utilization conflicts and forest degradation. The system of clear-cutting (with artificial regeneration) was adopted. Shiva concludes that converting natural forests to pulpwood plantations further eroded biomass needed for fuel, fodder, fertilizers etc. Monoculture plantations of eucalyptus, pine, etc., severely undermined ecological processes in mountain catchments like the Himalayas and the Western Ghats. They destabilized the hydrological balance of streams and rivers. In a discussion, Krishnaraj cites that India's sports needs fuelled the development of 'soft wood' cultivation of eucalyptus trees whose soft bark is used for making rackets.

Agarwal reviews the affect of environmental degradation on the gender-class system. Degradation mainly occurs in three areas – disappearing forests, deteriorating soil conditions and depleting water resources. Satellite data reveals that in 1985–87, 19.5 per cent of the country's geographical area was forested which declined at the rate of 1.3 million hectares per annum. By 1980, 56.6 per cent of India's land was suffering from environmental problems especially water and wind erosion. Flood occurrence increased during 1971–81. Citizen's Report documents that while India had 40 per cent land under forest, post-colonial and post-independence it has reduced to 12 per cent.

Impact on Fodder, Fuel and Water Availability

In India, many rural households gather essential items from village commons and forests. These items, including food, fuel, fodder, fibre, small timber, manure, bamboo, medicinal herbs, oils, resins, gum, honey and spices, are meant for everyday personal use and sale. Agarwal states that all rural households use the village commons to some degree. However, they are of critical significance for the poor. Data collected in early 1980 from twelve semi-arid districts of seven Indian states indicates that village commons account for at least 9 per cent of the total income for poor rural households (the landless and those with less than two hectares of land equivalent to drylands). In most cases it contributes to 20 per cent or more of the total income. However, for the non-poor, it contributes a mere 1 to 4 per cent. In contrast to self-sufficient landed households, the village commons supply more than 91 per cent of firewood and 69 per cent of grazing needs (Agarwal, 1992).

Roger Jeffery, Patricia Jeffery and Andrew Lynn's study reveals that presently more cow dung is used as fuel in Uttar Pradesh because of the shortage of scrubs and wood. According to their estimate, almost 30 per cent of the fuel needs of this area are met by cow dung which could have been used for composting or manure or for both fuel gas and manure through biogas technologies (Jeffery et al., 1989: 32–37).

Forests are a significant source of livelihood especially for tribals. Agarwal quotes N.S. Jodha who reveals that an estimated 30 million or more people in the country depend wholly or substantially on such forest produce for a livelihood. She makes us aware that foresting affects soils and the availability of ground and surface water for irrigation and drinking. Rich people tap groundwater by digging deeper wells or tubewells but the poor mainly depend on surface water.

Agarwal details the impact on women in terms of the effort spent in collecting fodder, fuel and water and the distance traversed in the process. She further studies the impact on income and nutrition. Health is another area of concern for women. There are direct links between environmental disturbances, illness and need for health care provision which is required due to increased morbidity.

On Time

Agarwal presents a table computed in Madhura Swaminathan's article in *Manushi*. Firewood is the single most important source of domestic energy in India (65 per cent in the hills and deserts of the North). Firewood collection time in the hills has increased dramatically in the last few years taking up to 4–5 hours daily after an approximate foraging over a distance of 5 kilometres. In the plains of Madhya Pradesh, it is sufficient to forage one to two times a week over a distance of 5 kilometres. In Rajasthan wood fuel collection requires travelling 5 hours a day (in winter) over 4 kilometres of land.

Drinking water shortages are characterized by traversing greater distances and waiting longer to fill vessels at a crowded source. Heavy irrigation increases salinity thereby aggravating the shortage of drinking water.

On Income

According to Agarwal, women's direct income is affected when the population gathers fewer items from the forest. Gathering takes more time during scarcity. Thus cultivation and crop production receive lesser attention, particularly in areas with significant male migration.

Shortage of grazing lands reduces fodder availability rendering goats and other milch animals beyond the resources of landless women. Vandana Shiva (1988) maintains that in the Garhwal Himalayas at least two-thirds of the fodder needs of farm animals were derived from the straw of cereal crops. Food and fodder sources were displaced when monetary considerations forced farmers to shift to market vegetables. New high breed crop varieties had shorter stalks and more grain which further compounded fodder shortage. Shiva states that commercial processes disturb the food cycle. A viable local economy is based on subsistence agriculture in which the woman is a soil builder and primary producer. Commercial processes drive women into becoming subsidiary earners and wage labourers. Women's traditional control over land use is displaced by the concept of 'ownership' where efficiency is defined as the maximization of 'profits' for the owner. Shiva claims that commercialization increases women's workload by pushing them to produce for the sake of survival. It also decreases the market value of their work. For example, selling firewood through head loading becomes difficult with increasing shortage of wood fuel as more time is spent in gathering the same amount.

On Nutrition

Diet is affected due to shortages in food and in gathering wood fuel. The scarcity of wood fuel induces people to consume raw or partially cooked food which could be toxic. According to Agarwal, studies conducted in Bangladesh indicate that the daily number of cooked meals in poor houses has declined. A trade off between time spent in fuel gathering versus cooking adversely affects the meals' nutritional value. Shiva points out that food abundance at the market level does not always indicate greater nutrition for the girl child. She cites C. Satyamala's study of Ludhiana district in Punjab. It reveals that within the same economic group, the percentage of undernourished female children was consistently higher than the percentage of male children of the same age. Amartya Sen's comparative study of two districts in Bengal (one where land reforms had taken place and commercial processes had started and another where no reforms took place) affirms that increased nutrition due to affluence is not evenly distributed among the children of both genders (Sen and Sengupta, 1983: 855–864). Krishnaraj points out that crop production has suffered due to the negligence of the government and the anti-agriculture bias in policy. The dominance of wheat and rice over 'coarse' cereals has increased people's consumption of these crops. Health standards are severely affected when nutrients are lost due to insufficient consumption of coarse foods (Krishnaraj, 2007: 43).

On Health

Apart from malnutrition, poorer people are more likely to be exposed to water polluted by pesticides and fertilizer run-off. This creates several chronic problems including water-borne diseases. Waterlogging causes many ailments which translates into women spending their time in nursing ill family members. Agarwal reveals that more than acceptable levels of DDT and BHC residues were found in the milk of nursing mothers among female agricultural workers in China. In India, pesticide exposure is frequently associated with visual disabilities.

Vandana Shiva (1988) charges that in commercial agriculture 'productivity' simply means making more money. This negates the productivity of women as food producers and processors. Commercial agriculture cripples the local economy and individual nutrition. Poisonous pesticides kill the fish in rice fields; reeds (for fibre and rope making) are destroyed by weedicides; rice and wheat bran (which are eaten when grains are processed at home) are destroyed by mechanized milling. Shiva is critical of commercial processes which destroy natural products and then put 'nutrition' back through the processing industry. Philippino women, who do 'barog' (a process of shaving off already beaten stalks of rice to glean the remaining grain), procure as high as 10 per cent of the total yield. The male scientists of International Rice Research Institute (IRRI) saw these women's gain as a 'loss' to the farmers and hence created a mechanized replacement. In India, Barbara Harris White calculated the displacement of women's labour due to the establishment of rice mills in Tamilnadu. The introduction of modern rice mills (under state patronage and foreign technical assistance through state subsidies), which competed with the hullers' traditional pounding system, affected female employment in the countryside (Harriss-White, 2007: 104).

Water Policy and Women

Shiva reminds us that rivers and water resources have been central to the prosperity and survival of the Indian civilization.

Resource-intensive utilization has led to a rapid disruption of essential ecological processes which recharge and renew water resources. Current water management focuses on tapping and distribution, irrespective of its affects on the hydrological cycle and without concern for natural replenishment.

During the period of technological euphoria, river diversions were assumed to lead to undiluted benefits. Today's age of ecological restraint is based on the bitter experiences of upstream and downstream destruction caused by large dams. The principle of equitable apportionment needs to be foregrounded in order to preserve the integrity of river basins and to minimize conflicts over water.

Contemporary water rights are the rights of a state to control or consume water using large water projects. Rules have been framed on water use to protect the rights of states to build dams. This framework assumes that a river is wasted if it is not dammed. It ignores the link between surface water and groundwater and between fresh water and marine life. It also forms the basis of legal approaches to rights and entitlements over vital natural resources. Water tribunals allow privatization and overexploitation of water resources in the basin by excluding control over groundwater utilization. This leads to acute water scarcity and drought proneness in the region. It further creates demands for inter-basin transfers and river diversions.

Shiva cites that the frequent occurrence and severity of drought in Maharashtra cannot be attributed to climatic changes alone. It is linked to the growing number of electric pumps which deplete the water table. Rapid groundwater depletion has increased the number of problem villages (i.e., those with no water source) to 23,000. In this area, sugarcane is the most water-consuming crop. It also increases salinity in the process. Nonetheless, with the help of the World Bank, vested political interests in Maharashtra are financing the expansion of sugarcane which has resulted in the decreased production of jowar and pulses.

The direct effects of this kind of water management are two-fold: (a) acute drinking water shortage resulting in women walking long distances in search of water, and (b) the displacement of people because of large dams. S. Parasuraman conducted a special study of the impact of displacement (due to development projects) on Indian women in 1993. Constitutional safeguards for SCs and tribals never find a place in resettlement plans. Only those who have formal access to resources are recognized as project-affected and thus compensated. The landless who depend upon the land for their livelihoods do not receive compensation or rehabilitation provisions. Nothing was done to facilitate their acquiring a new access to livelihood. Women face similar problems when they are forced to move. The displacement and resettlement process unsettles them more than other household members because they cannot articulate their needs and problems.

In the case of Sardar Sarovar, the women settled in Parveta village were initially happy because they got land. However, they soon saw significant changes in the quality of their life. Health dropped and mortality rose. Poor water quality and lack of green fodder caused cattle deaths. The growing need for money caused increased dependence upon cash crops. The agricultural inputs of women added considerably to their burden.

Parshuraman writes that the resettlement of the adivasi oustees (of Sardar Sarovar in Parveta) in the midst of villages inhabited by caste Hindus has intensified the 'Hinduization' process and high caste social practices. Tadavis have stopped taking brideprice and are moving towards dowry, *purdah* has become common, male children receive preferential treatment in schooling and parents demand reduced labour inputs from them. He feels that rehabilitation and resettlement programmes have strong biases toward men, such as land provision, subsistence allowance payments and government jobs (Parasuraman, 1993).

4

Overview of the Critique of Industrialism

> Industrialism is, I am afraid, going to be a curse for mankind. Exploitation of one nation by another cannot go on for all time. Industrialism depends entirely on your capacity to exploit foreign markets being open to you, and on the absence of competitors. – And why should I think of industrializing India to exploit other nations? Don't you see the tragedy of the situation, viz., that we can find work for our 300 million unemployed but England can find none for the three million and is faced with a problem that baffles the greatest intellects of England. The future of industrialism is dark. (Mahatma Gandhi, cited by Joshi, 2003)

This chapter critiques the industrial paradigm. It analyses the political economy of capitalism and its three pillars: the market, technology and environmental economics. The critique suggests that centralizing tendencies are inherent to the capitalist economy. Active decentralizing strategies and economic measures are required to counter these tendencies. There is a brief foray into the alternative economic and community life envisioned by environmentalists. Many view a decentralized economy with self-sufficient communities as ideal.

Not all the quoted authors are ecofeminists but their writings are frequently cited in environmental literature to point out the systemic problems of capitalist economy. These include Karl Polanyi (1997), Herman Daly and Robert Cobb (1990), Ted Trainer (1989), C.T. Kurien (1978), Peter Goering et al. (1993),

Marilyne Waring (1988) and Ivan Illich (1973). Environmentalists are critical of socialist economies (which collapsed after 1989) as they upheld industrialism just as much as capitalist economies. Industrialism is built from a belief in energy-intensive technologies and promotes the notion that happiness means possessing more material goods. A few strands within ecofeminism are critical of industrialism as a strategy of development. Some of the building blocks of ecofeminism have been stated explicitly or implicitly. The environmental critique (put forward by environmentalists) is presented briefly before proceeding to the specific contributions of ecofeminism.

Background to the Emergence of Eco-political Theories

In the 1960s widespread public concern over environmental degradation was voiced in developed countries. It was suggested that society lacked environmental planning. Marxists identified maldistribution of property and income as the primary cause. The publication of the first report called "Limits to Growth" by the Club of Rome in 1972 made people realize that the crisis had surpassed human technological ingenuity. State planning was impotent if the desired goal was an increase of material wealth. Thus began new eco-politics which is generally called 'green politics'. This also marked the beginning of the critique of industrialism.

The 1960s and 70s were marked by student movements all over Europe and America. They witnessed a questioning and revision of socialist theory. This period saw the rise of the New Left. Within the New Left, Herbert Marcuse, Theodor Adorno and Jurgen Habermas tried to trace the problems of industrial society, environmental dislocation and instrumental or technological rationality. They characterized the essence of the situation as a 'crisis of participation'. The concern for environmental equality was incorporated into the agenda (of greater individual and community autonomy and control) of the New Left. 'Decentralization' and 'self-management of power and resources' were seen as the ultimate solutions. Despite the radical and provocative nature of their critique, their concern was limited to the 'liberation' of 'inner' (and not 'outer') nature. Their political themes were grassroots democracy and social justice. Hence, they remained anthropocentric, i.e., people-centred. Their concern did not extend to placing non-human nature on the same plane (Eckersley, 1992). Around 1980, the French school led by Foucault was the most prominent critique of the modernization theory. Foucault made an attempt to dismantle and demystify positivist knowledge systems thereby inspiring environmentalists and ecofeminists.

In 1962, Rachel Carson's book *Silent Spring* brought the 'interconnectedness' of human beings with nature into popular consciousness. The book had many political repercussions and, according to Eckersley (1992), proved to be a theoretical breakthrough. A decade later, two reports were published – the aforementioned 'Limits to Growth' and 'Blueprint for Survival'. Both reports warned of an imminent 'crisis of survival'. They popularized the metaphor of our planet as a sinking spaceship. The general solution offered was to invest power in the government and to fight the crisis through environmental planning and technological fixes. These solutions overestimated the capabilities of centralized institutions. Conversely, they underestimated the potential of decentralized democratic and political institutions to respond to the crisis.

Eco-political theorists saw the emancipatory potential of the ecological critique of industrialism. They questioned the notion of material progress. Simultaneously, they demanded a re-evaluation of the foundations of and the conditions for human autonomy (or self-determination) in Western political thought. Murrey Bookchin pointed out that "the cultivation of ecological society, resting on the principles of social ecology, will serve to expand rather than narrow the realm of freedom or self-directedness in first (i.e., non-human) and second (i.e., human) nature" (Eckersley, 1992: 20). Eckserley describes these attempts to widen the agenda of emancipatory eco-political theory on three levels – human needs, technology and self-image. She treats it as a synthesis of the working class movement (which tackled authority by focussing on the issues of participation and distribution) and the ecological movement (which questioned the basis and structure of needs). Another concern is cultural renewal including the revitalization of civil society. An attempt has been made to integrate issues raised by the other social movements such as feminism, peace, Third World aid and development. The writings of the New Left on the issue of social ecology have not been included in this work.

Environmental Critique of the Political Economy of Capitalism

Economic Growth

Capitalist economy rests on constant economic growth. Many liberal economists object to equating development with economic growth. They want to add human resource development, health development etc., as special concerns and believe that the market can be regulated by the state. Furthermore, international agencies can bring about structural changes by pressurizing local governments. Thus, they suggest that development is possible through redistribution within the capitalist economic framework. Environmentalists criticize this position strongly by calling it unrealistic. They propose a paradigm called "say no to economic growth within development".

Three key factors control the current dynamics of economic growth which relies on the underdevelopment of Third World countries making structural changes for redistributive justice within such a country impossible or, in other words, are responsible for inequality and destruction of nature. These are:

(i) The market, particularly international trade, is guilty of polarizing development and the appropriation of natural resources by the rich in both the North and the South.

(ii) Non-renewable energy-based technology development which forms the basis of the industrial mode of production, is accelerated through market dynamics and has an immense destructive impact on the ecological balance of our planet.

(iii) Environmental economics, which was devised to make producers pay for environmental degradation, tries to internalize 'externalities' but does not bother about lost ecological values that cannot be restored.

The Market

The market is the key to understanding 'underdevelopment' as it gives direction to the present global economy. This economy allows access to scarce resources to those who can pay the market price. Such consumers can also influence investment decisions. Resources are developed on the basis of the profit maximization principle. Only products with high profit margins are produced because the rich can pay steep prices. Thus product development is indiscriminate. Since assets and property are unevenly distributed from the beginning itself, product development becomes inextricably linked with market mechanisms. Therefore, the just distribution of resources should not depend on the market. The economic historian Karl Polanyi describes social development associated with the rise of the market as the 'Satanic Mill'. His works of 1944 begins with the sentence: "At the heart of the industrial revolution of the eighteenth century was an almost miraculous improvement in the tools of production, which was accomplished by a catastrophic dislocation of the lives of common people" (Polanyi, 1944, 1997: 33). Before the emergence of the market, trade was done through barter system or through itinerant sellers.

Maldistribution of resources and wealth occurs because rich countries outbid poor nations and the rich within a country outbid the poor of that country. The global market economy delivers resources to a few thereby depriving the majority. Maximizing output, sale and investment returns directs productive capacity towards the rich. Thus, a high economic growth rate accelerates the very mechanism which causes the problem in the first place. Global markets allow the interior of the Third World to produce for the market and enter the cash economy. However, these people lack bargaining power and thus earn lesser. The terms of trade (i.e., the price at which poor countries sell to and buy from the rich) are pitted against the poor. Rich countries form a "buyers' market" to dictate prices (monopsony). Even though poor people form the majority, they do not enjoy absolute control over their products (monopoly). This forces them into monoculture of cash crops thereby increasing their vulnerability to the fluctuating prices of food in the market (Trainer, 1989; Shiva, 1988; Goodland and Daly, 1992).

The productivity of the Third World has been moulded to suit the demands of developed countries. The market pulls people away from subsistence production and forces them into the cash economy. It promises modern amenities but destroys their self-sufficiency. Development dismantles and disintegrates while promising to reconstitute – a hollow assurance which is never realized in the Third World (Trainer, 1989: 77).

Textbooks on international trade theory acknowledge that gains are distributed unevenly between countries or trading blocks depending on their economic clout and their disposition to flex it. The theory does not identify a 'process' which could equalize the advantages/disadvantages of countries participating in international trade. Trade theory postulates 'comparative cost' – for example, a product X can be produced by two countries A and B. However, what matters is which country can produce it more cheaply. Comparative cost theory does not explain 'competitiveness' which is a matter of market clout. The concern for growth obscures sustainability and redistribution. In the Third World the utilization of natural resources is influenced, directly or indirectly, by the trading partners, i.e., developed countries. Sustainable practices in the South are disposed in favour of closer economic dependence on technology and markets of the North. The environmental effects of liberalizing trade and stimulating agricultural exports are often harmful. Environmental degradation will continue to be propagated until attention is diverted towards the ecological and cultural systems which convert natural resources into livelihoods (Radcliff, 1987). Utsa Patnaik charted the shift in the land-use pattern of Third World countries (particularly India and Mexico) in response to the needs of industrial powers (Patnaik, 1993, 1996). Regrettably, today, we grow roses and tulips on our scarce land instead of more food. Foreign investment, debt and aid force the surplus back to rich countries (Trainer, 1989; Radcliff, 1987).

Traditionally, women were responsible for non-mechanized labour-intensive production, food processing and family nutrition. Men displaced women in the commercial agriculture of cash crops (Radcliff, 1987; Agarwal, 1992). Census data shows a feminization of the rural workforce but an increase in casual labour. In 1991 female workers constituted 32 per cent of the rural workforce which increased to 36 per cent in 2001. Such tendencies were particularly pronounced in the 150 poorest districts which were selected for the first phase of the National Rural Employment Guarantee Scheme. This is because of their lack of geographical mobility and an inability to reject socially prescribed familial responsibilities (as mothers and care givers) which absorb a major part of their life (Krishnaraj and Kanchi, 2008: 47–48).

Instead of pursuing a policy of sustainable development, particularly for resources controlled by women, successive governments have enforced a survival strategy on rural households. This strategy involves intensive cultivation, capital stock depletion and migratory patterns which increase cash income (Radcliff, 1988: 56–78).

C.T. Kurien shows that colonization in the Third World accelerated the process of transforming merchant capital into industrial capital because the Money-Commodity-Money (Capital) (M-C-M) circuit came from elsewhere (colonial countries). Since the M-C-M circuit expands when transport facilities and communication links are provided, there naturally evolved a single national market responding to the purchasing power of any part of the country. This undermined locally related production processes of people involved in the Commodity-Money-Commodity (C-M-C) circuit (i.e., those producing commodities for local exchange). Consequently, production priorities, local markets and livelihoods get distorted (Kurien, 1992: 270–77).

The current policy of Special Economic Zones (SEZs) illustrates the pushing of industrialism at the expense of agriculture and people's survival concerns (see Annexure 1). Krishnaraj asserts that the policies of industrialization (such as buying food from outside) which worked in Japan and China will not work here. We have vast tracts of arable land, varied eco-climatic regions and a striking biodiversity. She accepts the critics' remarks that she is an agricultural ideologue (Krishnaraj, 2008).

The Market Abstracts from the Community and Creates Externalities

Daly and Cobb (1990) analyse the market in detail. Though they favour market institutions, and do not advocate a command economy but are concerned about the self-effacing tendency of competition in laissez-faire economies. Thus, either a limit over monopolies has to be imposed by the government or the scale of the production unit has to be such that it only serves regional markets. Daly and Cobbs believe that the government should encourage regional economies.

Even though the market is supposed to be driven by the self-interest of individuals (as assumed by classical economists), there is a need for consensus over some values like honesty, freedom, initiative, thrift etc. The market's dependence upon the community is two-fold: the latter regenerates 'moral capital' and 'natural capital'. The major problem lied in the existence of public goods and 'externalities' while the major challenge is the internalization of these externalities. They define externality as occurring "when production or consumption by one firm or consumer directly affects the welfare of another firm or consumer where directly means that the effect is not mediated through any market and is consequently unpriced" (Daly and Cobb, 1990: 53).

Daly and Cobb highlight the limitations of the concept of Gross National Product (GNP) as a tool for measuring growth. Economists measure the success of a particular sector or of the entire market using the GNP. A healthy market is characterized by a growth in GNP or per capita GNP. However, GNP only measures certain aspects of welfare. Typically, the whole is divided into parts which are studied. It is assumed that when the parts are added together again they have remained unchanged by their abstraction from the whole. Recent criticism has pointed out that the growth of GNP has been at the cost of psychological, sociological and ecological terms. Daly and Cobb forget to mention that this cost includes the unpaid labour of women within families and communities.

Daly and Cobb further reveal that the state manipulated the markets under the ethos of rising industrialism. They analysed the history of conflict between landlords and industrialists and concluded that when agriculture was capitalized the government artificially reduced the prices of food and other resources to ease the conflict between capitalists and labour. They feel that land (specifically the biosphere) is neglected because it never demands its own 'scarcity' price. They remind us that unlike capital and labour, land cannot be reproduced.

Polyani avers that although the self-regulated market is supposed to be a key institution of the capitalist economy, it is frequently manipulated in favour of industrialists to accelerate industrialization (Polanyi, 1944, 1997: 75–76).

Technology

Industrial production depends upon non-renewable energy resources for the exploitation of natural resources at an accelerated rate. The factory system (inherent to industrial production) commodifies labour. This process is fundamental to both market systems and command economies as the ultimate goal is to accumulate. The World Market Economic System sets the rules and regulates. Hence competing state economies continue developing technologies while disregarding their consequences. Saral Sarkar (1983) has documented the tremendous destruction of natural resources during the socialist regime in Russia. He claims that this was a crucial factor in the collapse of the socialist state.

Technology affects society in two ways: (a) ecological destruction which deprives large masses of their livelihoods, and (b) a social ecology crisis because of the institutional requirements of these technologies. Vandana Shiva articulates the paradoxical impact of technology by stating that an increasing dependence of industrialized societies on natural resources is accompanied by the widespread myth that dependence on technology implies a decreased dependence on nature and natural resources. The conflicts become visible when resource and energy-intensive industrial technologies are challenged by communities whose survival depends upon the conservation of resources (Shiva, 1991: 13).

Technology Helps the Market to Expand

Transport technology is energy intensive and facilitates market expansion thereby disturbing local resource allocation (Kurien, 1992). In a market economy, ecological limits are generally invisible as most resource intensive technologies operate within enclaves removed from the resources they operate on. Vandana Shiva blames energy intensive long distance transportation technology as a part of the process of development (Shiva, 1991: 35).

Efficiency is Derived from External Energy Sources

Transportation technologies aid the development of larger markets (which in turn increase the need for transportation technology). The technology choice is justified in contemporary market economies on the basis of 'productivity' and 'efficiency'. Productivity is judged in terms of 'productivity per labourer' and not as 'productivity per unit of natural resource'. Increased consumption of non-renewable natural resources contributes to the increased productivity of a modern farmer or a factory worker. Shiva reveals the fallacy of the argument of 'productivity' by deploying an estimate (provided by Lovins) of the energy needs of the world's population. Currently, man's annual global energy conversion from all sources (wood, fossil fuel, hydraulic power, nuclear) is approximately 8x10 watts per person. This is more than twenty times the energy content of the food necessary to feed the present world population (at the FAO standard of 3,600 calories per capita per day). Each person on earth, on an average, possesses the equivalent of about fifty slaves each of whom work for forty hours a week. There are enormous inequalities in the distribution of this 'slave' labour. The average inhabitant of America, for example, has 250 times the number of slaves as the average Nigerian. This forms a substantial reason for the efficiency difference between the American and Nigerian (Shiva, 1993: 36).

According to Odum (cited by Shiva, 1993), efficient livestock management for the production of meat requires 10 calories of energy input to produce one calorie of food energy. Nearly a century ago in the West, one calorie of food was produced by using a fraction of a calorie of energy input. The subsidy provided by the capital stock of the earth's non-renewable resources is the reason behind the invisibility of this high energy cost of the resource-inefficient method. Subsidies can make counter-productive processes appear efficient. For instance, the European Union gives 40 per cent subsidy for agriculture and hence sells its dairy products at a cheap price to the Third World.

Vandana Shiva claims that when the production from high yielding varieties of seeds is evaluated on the basis of productivity per unit volume of water input, these miracle seeds prove to be two to three fold less efficient in food production than, say, millet. She concludes that the hallmark of modern technology is reduced labour as opposed to reduced energy (Shiva, 1993: 37).

Industrialized Agriculture: Last Straw on the Planet's Back

Technological advances in the field of agriculture are attacked vociferously because they accelerate the negative impact on the very natural resources which provide food security in the first place. Peter Goering, Helena Norberg-Hodge and John Page analyse the main characteristics of industrial agriculture (which wreaks havoc on the food chain) in the following way (Goering et al., 1993: 3–6):

1. *Specialization:* Contemporary agricultural scientists work on extremely specialized fields of knowledge. For example, research on the design of new pesticides will typically be focused almost exclusively on the 'target' pest. Questions

about the impact of the pesticides on other (potentially beneficial insects) or on human health (through chemical residues in food or drinking water) are often ignored. The specialization extends to farmers. They are often unable to return to more diversified practices as they have invested heavily in specialized equipment.

2. *Standardization:* Standardization is a critical factor in the modern drive towards economic efficiency. Thousands of regionally-adapted species and varieties have been driven to extinction because they failed to meet the narrow requirements of the industrial food production system.

3. *Centralization:* The economies of scale and the growing transportation and marketing networks prefer centralization. Today, the GATT (General Agreement on Tariffs and Trade) and the World Trade Organization (WTO) threaten to extend these networks still further. This will eventually concentrate power over the whole global market in the hands of a few agri-businesses.

4. *Equation of animals with machines:* Animal production has become highly centralized. It is treated as an industrial product with the intended goal of fattening it as quickly as possible. As a result, one-third of the world's grain production is now fed to animals (especially in rich countries). Factory farms collect animal manure from feedlots. However, this manure cannot be used because of its high salt content. The runoff from the feedlots pollutes the atmosphere and both surface and groundwater.

5. *Mechanization:* Mechanization goes hand in hand with large irrigation schemes. Both processes degrade the soil either through evaporation of the soil moisture (in dry land) or by increasing salination and waterlogging (in irrigated land). The use of fossil fuels, fertilizers and pesticides has increased due to mechanization. This pattern of production will eventually deplete our natural resources.

Environmental Economics

Environmental economics treats the environment as an extension of a human right – a "common property". It works on the principle of compensation. Many economists advocate the principles of 'internalizing externalities' and 'environmental management'. Hazel Henderson warns that the current crop of narrowly trained economists use economic principles, such as 'willingness to pay for clean air' or 'to keep it unused' etc., as a part of environmental management which legitimizes polluting activities. They do not take the future into consideration (Henderson, 1981: 231).[9]

Quantity Emphasized but Quality Ignored

A special issue of the *Ecologist* (1992) on development made a scathing attack against cost-benefit analysis which reduces everything to money value without factoring the quality of the result. Some values cannot be commensurate with others without being lost entirely. Measuring costs against benefits sets centralized global management in motion which is absurd from an eco-political standpoint. The *Ecologist* fears that Ivan Illich's forecast about the coming steady state society being an oligarchic, undemocratic and authoritarian expertocracy will come true.

Marilyn Waring has examined the proposals for computation and imputation of value to environmental destruction and regeneration. In the words of Schumacher, "In fact, however, it is a procedure by which the higher is reduced to the level of the lower and the priceless is given a price. It can therefore never serve to clarify the situation and lead to an enlightened decision. All it can do is lead to self deception or to the deception of others; for to undertake to measure the immeasurable is absurd" (cited in Waring, 1988: 297–98).

Women's Work Made Invisible

Feminists criticize classical and neo-classical economic theory of making women's work invisible. The problem exists on different planes – from the formal level of enumeration (because of the biases of the enumerators and the enumerated) through the level of planning and finally at the conceptual level. According to Marxist feminists, the question exists with regard to women producing use values or exchange values. In 1988, Maryline Waring exposed the 'misplaced concreteness' of the concept of GNP in the United Nations System of National Accounts (UNSNA). She criticizes the concept of 'production boundary' in UNSNA which decides what is considered production and productive labour. Although amendments have been made over the years, the major stumbling block is the definition of household activities, which include collecting fodder, fuel and water, producing vegetables in a kitchen garden, cooking, babysitting, nursing the old etc. She claims that "... so long as man is counted head of the household, it is immaterial how much efforts are made to measure women's household production for self consumption or also for the market, all the labour is counted under his name. The rising phenomenon of female headed households in the developing countries has not made any dent in the concept of head of the household."

Many agricultural operations performed by women are not included in the questionnaire which is used to collect data for the UNSNA. Statisticians rue the fact that the diversity of women's situation does not allow them enough space to prepare a proper slot/label for women's work or household activities. They claim that women's work cannot be universalized and hence cannot be quantified. This need for 'enforced consistency' leaves women's unpaid activities unacknowledged.

It is difficult to impute the cost of activities which do not come into the market, particularly in the Third World (because comparisons of opportunity cost or compensatory cost are unavailable). Marilyne Waring responds to this difficulty by citing a passage from a text by Raj Krishna, an Indian economist, who was a consultant to the Asian and Pacific Development Council. He was confident that the cost of activities could be easily computed in the case of quasi-productive work as it was paid for in the commercialized sub-sector/market. He states that studies could be sponsored using this practice to collect imputation-related data such as time allocation and wages. He argues that many national products are imputed by national accounts statisticians using similar procedures (also mentioned in Krishnaraj, 2008).

Waring faces a dilemma about whether to demand imputation of women's unpaid work. Although such data would be invaluable to feminist policymakers, it is a reformist demand which accepts the basic principles of the current system. This policy could reduce women's lives to a mathematical formula.

Need for Decentralization

Despite their ideological differences, both capitalism and socialism rely heavily on non-renewable resources. They tend to exploit renewable resources and the waste absorption capacity of the environment at unsustainable rates. Both are committed to large-scale, factory-style specialized production units which are energy and capital intensive and hierarchically managed. The industrial mode of production requires a large, centralized and bureaucratic organizational structure. The market itself brings about large-scale dislocation, dismantles communities, disintegrates families and encourages the individualization processes. Such atomized individuals are controlled and managed by a bureaucratic state, centralized health care system and information industry thereby causing a loss of cultural identity and their sense of being a part of the civil society.

Many eco-philosophical streams contribute to the overarching discourse of community versus state control or community versus market control. There are two prominent perspectives within which this discourse takes place: the ecocentric and the anthropocentric. The ecocentric perspective regards the question of our proper place within nature as logically prior to the appropriate social and political arrangement of human communities. Its ethical position regards the various parts of the biotic community as valuable for their own sake. Anthropocentric emancipatory theorists, on the other hand, maintain a greater continuity with the New Left in raising the issues of participation and survival along with the crisis of culture and character. However, both streams have critiqued economic growth and technology and are against large-scale organizations. Both reject the free market which is responsible for the loss of commons such as water, grazing grounds and forests. To some extent, the Marxian theme of self-realization (actualization of potential of self through an equal access to resources) still influences the anthropocentric view.

The general theme of the ecocentric stream is that industrialism presupposes a universal and virtually unequivocal acceptance of industrial growth and expansion. In such circumstances, nature is regarded as external to humans and defined as a resource to be exploited for human gain. This attitude characterizes industrial culture as a whole and reflects the dominance of material value over aesthetic, spiritual and other human values. Keeping this in mind, ecocentric environmentalists favour smaller communities. The anthropocentric view believes in smaller communities as well. However, it has not arrived at a clear conclusion against industrialism.

The Buddhist world view regards humans, other living, non-living systems as internally connected. In fact, their notion of 'Self' is not an atomized individual but an 'inter-being' (Keown, 1996). Damien Koewn says, "The universe, therefore, comes to be seen not as collections of more or less static but dynamic net work of inter related causes and effects."

Smaller Communities: Less Alienation

Ivan Illich articulated the need of smaller communities by asserting that industrialized countries tend towards centralization on one hand and social atomization on the other (while industrialism persists). Industrialism is organized for indefinite expansion and the concurrent unlimited creation of new needs. The concern for 'intimacy of scale' is ruptured when specialized institutions replace family and community. Illich's vision of an alternative 'convivial society' is based on the preconditions of clear limits to institutional growth and the search for natural scales of organization. He envisions, "It will be such a society, in which modern technologies serve politically interrelated individuals rather than managers" (Illich, 1973, as cited by Jones, 1990: 112).

Deep ecology shares a similar concern about the destruction of communal life. Overconsumption and militarism are seen as the fundamental causes of the degradation of nature and the suffering of the Third World. Within the human centred (anthropocentric) social perspective of critical theory, overconsumption is seen as a social relationship or a problem of ownership (misdistribution between the wealthy and the poor). However, within the ecocentric perspective, non-human nature is seen as a community and valued in its own right. Hence human overconsumption is considered wrong.

The bio-regional movement in the US was a consequence of deep ecology. It articulated an alternative vision of human society which would be more humane and which would offer a tenable symbiosis between itself and the rest of nature. This involves a radical decentralization of human society at political and economic levels. Thus genuine democracy and local control of economic activity become possible. The advocates of bio-regionalism believe that political and economic decentralization would make humans more ecologically rational forms of life.

Their description of the advantages of decentralizations can be summarized as follows:

— The identification of place is one basis for an expanded identification with all forms of life.

— It will give a chance to continue cultural evolution.

— Decentralization, by shortening the feedback loops between economic choice and consequence, facilitates social learning and the possibility of ecological wisdom (McLaughlin, 1993).

The decentralization principle within the concept of deep ecological thought coincides with the feminist concern for combining women's reproductive role with their productive role to initiate participation in political decision-making. The latter can be realized under a decentralized economic and political system. Although feminists express certain differences regarding its core philosophy, Deep Ecology explicitly mentions the parallels with the dominating systems in sexism. They posit the ecocentric view thereby elevating nature and nature like phenomena. Bio-regionalism indirectly contributes to the feminist vision of an alternative society by stressing the need for devolution of political and economic power. Even Gandhi's idea of a 'village republic' conflates with the decentralization.

Gandhian Critique of Industrialism

Gandhi critiqued industrialism through his lectures and writings. He boldly analysed that the process of industrialization required incessant surplus accumulation which led to the colonization of other nations. His analysis is concerned with finding employment for the youth of India who left their traditional occupations in rural areas to look for jobs in urban factories. For him, the term 'industrialism' implied an obsession with mechanized 'mass production' which displaced labour. He was worried about the farmers who could only work for four months each year in the kharif season. His solution was the spinning wheel because of its potential for creating employment. He was aware that for any production activity a machine would be required. He favoured simple and elementary machines which could be provided to every house. The labour of each person could be converted into grain and cloth which could be exchanged without the mediation of international markets. He fretted over the use of non-human energy for operating big machines. It appears that he was aware of the basic principle of ecological science which advocates that we should leave behind very few ecological footprints on this earth.

Gandhi was not totally against imports. His quote from the 22nd May, 1927 issue of *Navjeevan* clarifies, "The general principle is that we should produce or make whatever we can in the villages themselves .... If however, we cannot make the sewing needle in the village, we should not set our face against the easily available and cheap needle from Austria. I see nothing wrong in accepting from any quarter whatever is good and worth accepting and which we can assimilate" (cited in Joshi, 2003: 6). He was against aping the Western model of urbanization which suited only small countries like England. He was aware that what was good for one nation in a particular situation would not necessarily translate as being advantageous for another. He was willing to introduce machinery only where it was absolutely necessary (such as while producing life saving appliances).

The Gandhian explanation resonates with Rosa Luxemburg's analysis of capitalism (as cited by Mies et al., 1988). According to Luxemburg, the world capitalist system expands through accumulating surplus by the appropriation of value created in the non-capitalist economies of the world. It is impossible to build capitalism in a country by simply accumulating surplus (required for huge expansion) through the exploitation of the working class of the same country. Mass production requires both raw materials and markets from non-capitalist countries. Thus capitalism thrives on the ruins of non-capitalist economies. Mies and her colleagues use Luxemburg's thesis to formulate a strategy of disengagement from the world capitalist system. Thus, they arrive at the subsistence perspective.

The French anthropologist C. Meillasoux has mapped the surplus accumulation process from Turkey to Germany. Migrant workers raised by small farmers, especially their wives who are part of the subsistence economy, are employed in German factories and cities. The cost of raising, educating and training them is not incurred by the German economy. Their labour is cheaper than that of German workers. Hence, they contribute to a better flow of surplus value to German economy (Meillasoux, 1981).

Marxist and Dalit activists in India raise a contradictory note by stating that the hierarchically ordered caste system will never allow any village to develop into an organic, homogenous community. They claim that the urbanization and proletarianization process is essential for integrating castes into mainstream Indian society. However, I feel that an older community need not be glorified. The elements which sustained it can be culled to create a new community.

Annexure 1
Special Economic Zones (SEZs) in India

The Special Economic Zones Act, 2005 was passed by the Parliament in May 2005. It received Presidential assent on 23rd June, 2005. The SEZ Act, 2005 (supported by SEZ rules) came into effect on 10th February, 2006. Its main objectives are:

(a) generating additional economic activity;

(b) promoting exports of goods and services;

(c) promoting investment from domestic and foreign sources;

(d) creating employment opportunities; and

(e) developing infrastructure facilities;

As per the SEZ Act of 2005, SEZs are geographical regions which have different economic laws than a country's typical economic laws. The most lucrative incentives to woo investors include 100 per cent tax exemption for industries on export profits for the first five years and the provision of 100 per cent direct foreign investment in industries under automatic route.

The benefits of SEZs include:

— They offer economic progress to an area, its local inhabitants and the country as a whole.

— Exemption from payment of stamp duty and registration fees on the lease/license of plots to the SEZ developer.

— External commercial borrowings of upto 500 million USD a year without any restriction of maturity to the SEZ developers.

— SEZs and related projects have instigated certain controversies and a debate regarding their disadvantages.

— Revenue losses due to various tax exemptions and tax benefits awarded to Indian SEZs.

— Indian real estate developers are interested in setting up SEZs to take advantage of the Indian real estate bubble as SEZ land can be acquired at cheaper rates to create a land bank.

— All infrastructure cost is borne by the state.

Issues involved

- SEZs need land to build massive infrastructural facilities. The only contiguous land which is easily available and connected to the mainland is fertile agricultural land. Both the centre and the states eagerly wish to establish SEZs on this land to boost economic growth.
- There is widespread apprehension that SEZs will be exploited by developers who, with governmental help, will corner huge stretches of rich, agricultural land by giving a measly compensation to farmers without any resettlement and rehabilitation policy.
- A rough plan is that 40 per cent of the SEZ will be allocated for greenery, sewage and water treatment, 25 per cent for real estate development and the remaining 35 per cent for SEZ development. The Finance Ministry is apprehensive that they will lose revenue to the tune of Rs 1,00,000 crore annually once the zones are up and running. With its huge fiscal deficit, India can hardly afford the loss.
- Many SEZs are IT-related software parks and technology hubs. Critics argue that there is no need for setting up a SEZ for the IT sector. It has anyway done well in India as we have a talented and burgeoning human resource. The SEZ was an ideal dream for IT companies as the tax sops given to them are about to expire within four years. If they had built an SEZ within that time, they could have benefited from tax exemptions for another decade or more.
- Protests by farmers and activists (in various parts of India) prompted the government to declare that agricultural land would not be taken away to build SEZs. For example, farmers near Mumbai and Pune are angry that the land sharks of SEZ developers are targeting their land.
- There is a distinct possibility that many existing manufacturing units will shift base to the SEZ to escape tax. A SEZ would make their profits soar by providing a great place to do business and to reap huge tax benefits.
- The SEZs will mainly come up in Maharashtra, Gujarat, Uttar Pradesh, Tamil Nadu, Orissa, Karnataka and Haryana. Many underdeveloped states will continue to remain so. This could lead to regional imbalances which will throw up complicated problems.

No SEZ, for example, will come up in the north-east which already suffers from alienation and discrimination.

- The Commerce Ministry is pushing the argument that farmers will be well compensated. But from earlier experiences we know that farmers are not aware of properly handling liquid cash. They have never seen this kind of money before. As they do not know how to manage, invest or recycle it, they waste it in gambling or drinking. Farmers are never counselled on how to spend money or how to make it grow. The money is spent soon after the land is sold. Having lost the only livelihood they knew, they get caught in a whirlpool of debt.
- Another argument is the rehabilitation of farmers in the zones with jobs. Most SEZs are going to be IT related. Farmers can not be expected to do a nine to six desk job with machines they do not understand. These issues do not have immediate and easy solutions.
- Our food security will be affected when thousands of acres of agricultural land are converted into concrete jungles. It is highly likely that the farmers who sell their land will not get a fair deal. Such land acquisition will undoubtedly cause environmental degradation.

5

Ecological History

Relationship between Nature and Human Beings (Changing Resource Use) and between Men and Women

> If we keep in mind that 'productivity' means that specific capacity of human beings to produce and reproduce life in a historic process then for our further analysis we formulate the thesis that female productivity is the precondition of male productivity and of all further world-historic development. This statement has a timeless dimension as well as a historic one. (Mies, 1988)

This chapter traces the historical shaping of the relationship between nature and humans. Ecologists focus their investigations on the industrial revolution which was a significant technological leap for humankind. However, ecological history is rooted in the beginning of civilization when the hierarchical structure of society (based on class and gender) was established. The industrial revolution replaced the previous system of entitlements and access to natural resources. It paved the way for technology and market institutions to exploit natural resources. The instrumentalist approach to nature is not entirely new. This approach explains that the patriarchal relationship between women and men is based on the sexual division of labour. In early patriarchal relations, women's activities of generational and domestic reproduction were naturalized. The ideology of domination (which equates women with nature and controls both) emerged with the establishment of patriarchy. Tracing its history explains how women and men had different relationships with nature since the very beginning. This was not based on biology alone but also depended on the activities that they pursued for survival.

The arguments of four authors – two of whom are considered ecofeminists – are illustrated here. Maria Mies (1986 and 1988) traces the development of the instrumentalist approach (deployed by men) to the pastoral period. Carolyn Merchant (1983 and 1995) critically inquires into the ideological developments of the industrial revolution. She elaborates on the attempts at justifying violence against nature and women using scientific rationality. New technology was promoted using a new scientific framework and its underlying rationality. Merchant focuses on how the same process transformed the status of women under patriarchy. In *Death of Nature* (1983) she describes the vast geographical changes that Europe underwent since the beginning of the industrial revolution. Energy-intensive technology developed during this period enabled mastery over nature.

The third argument is jointly presented by two Indians who do not refer to the role of women but integrate an environmental aspect into Marxian production analysis. Marx's writings have inspired the feminist movement. Traditionally, production analysis provides a framework to understand the dynamics of society (including that of the changing status of women). In contemporary society, nature is perceived as necessary raw material. Even Engels acknowledged the devastating impact of industrial activities on nature in his book *Dialectics of Nature* but did not further develop the analysis. Technology plays a positive role in traditional Marxist analysis as the emancipation of humans is conceptualized as consequential to the development of productive forces. Ramchandra Guha and Madhav Gadgil (1993) categorize the relationships of humans with nature (according to the individual modes of production) into gathering, nomadic pastoralism, settled cultivation and industry. They establish close links between technology utilization, social organization and resource use.

Search for Origin: Finding a Materialist Concept of the Human Being

Maria Mies' early writing was a search for the causes of the pre-historical subordination of women. These writings help us to take a biological view in order to investigate how certain activities became the basis of a materialistic conception of women and men. The identification and glorification of women and nature in the mainstream symbolic system of the past is in stark contrast to the present derogatory connection. This shift cannot be simply rejected as mystification or sheer manipulation. The link between nature and women has a tangible, materialist basis. However, its interpretation is ideological and need-based.

In this light, the foundation of the asymmetrical relationship between men and women can also be explored (albeit in a conjectural manner). Mies quotes Leukert while stating that tracing the beginnings of human history does not involve fixing a certain date. Rather, it raises the question of finding a materialist conception of human beings and history (Mies et al., 1988: 70).

Eco-political streams developed ecocentric philosophies (which claimed to be more holistic and unified) to counter the anthropocentrism of the rationalist tradition. Postmodern feminism and ecofeminism attempt to point out the dualism between the sexes. However, neither of them forms a materialistic (and not biologistic) conception of the men-women relationship. Such a notion would be based on their activities and not on the current 'identities' of masculine and feminine. Mies' paper "The Social Origin of the Sexual Division of Labour" presents this argument (Mies, 1988: 67–95). Gerda Lerner (1986) has also traced the history of origin of patriarchy but does not theorize that the sexual division of labour had materialistic basis.

Mies insists that studies in primatology, prehistory and archaeology are useful for our search. However, they cannot be expected to explain "the worldwide historic defeat of the female sex" (Engels, 1973: 34). This will ring true unless we can develop non-biologistic concepts of men and women and their relations to nature and history. She iterates that the definitions of key concepts are 'occupied' (like territories or colonies) by the dominant sexist ideology. Her search for materialistic concepts of men, women and history is an attempt to redefine these key concepts like productive labour, production of life etc.

Human nature evolved historically through social interaction along with an interaction with nature. It cannot be understood if we separate its physiology from its history. Here, Mies transcends both the extreme positions of constructivism (postmodern feminism) and essentialism (ecofeminism). She grounds her analysis in history and deconstructs the human-nature relationship.

Mies suggests that the difference between human beings and other animals is that animals live while human beings produce their lives through work. She quotes the three premises of Marx and Engels: (a) people must live to be able to make history; they must produce things to satisfy their needs, i.e., food, clothing and shelter; (b) the satisfaction of needs leads to new needs; human beings develop new instruments to satisfy their needs; and (c) men who reproduce their daily life must make other men, i.e., they must procreate (the relationship between men and women). Marx characterizes the 'production of life' as a "process of appropriation of nature" through sensuous and objective activity (Marx and Engels in *The German Ideology*, 1970, as cited in Mies et al., 1988).

The Relationship of Men and Women with Nature Evolved Differently

Mies grounds her search on the definition of 'work' stated by Marx in *Kapital*. Work is the appropriation of the environment in which men and women develop and change external nature. Marx did not take into account the historical fact that men and women appropriate nature differently. Here, Mies employs a dialectical materialist conception of history (which remained a blind spot with Marx). The difference is usually obscured because 'humanness' is identified with 'maleness'. Mies quotes Leukert who states that this has led to "two qualitatively different forms of appropriation of external nature, i.e., two qualitatively distinct forms of relations to the objects of appropriation, the objects of sensuous bodily activity" (cited in Mies et al., 1988: 73).

Mies summarizes the development of men's domination over women and nature in the following way:

(i) Their interaction with nature (internal and external) was a reciprocal process. They conceived their own bodies as being productive in the same way that they conceived external nature as being productive.

(ii) Though they appropriated nature, this relationship did not constitute a relationship of dominance or property relation. They were not owners of their bodies or of the earth, but cooperated with their bodies and with the earth in order "to let grow and to make grow".

(iii) As producers of life, they became the first subsistence producers and the inventors of the first productive economy, social production and social relations (i.e., of society and history).

Mies searches for a materialistic dimension of the male body against this background. Her main thesis is that the male body has to mediate with external nature (through tools) to feel the productivity of its own body. The male's self-conception of being human (i.e., a productive being) is closely linked to the invention of tools and the control of technology. The same understanding is used to trace the history of mankind. She asserts that female productivity is the precondition of male productivity and of all further world historic development.

This materialistic understanding is employed in tracing the history of the asymmetrical relationship between men and women and the hierarchical development of the sexual division of labour. She provides evidence to prove that the survival of mankind has depended more on 'woman-the-gatherer' than on 'man-the-hunter'. This is in contrast to the assertions of old and new Social Darwinists.

Even if we demystify the 'man-the-hunter' hypothesis and assume that the great hunters would have been unable to survive without the daily subsistence production of women, we still cannot answer the question that why did not women (in spite of their superior economic productivity as gatherers and early agriculturists) prevent the establishment of a hierarchical and exploitative relationship between the sexes. Mies explains this in the following way:

(a) The hunter's main tools are instruments which destroy (and not produce) life. They are a means of destruction.

(b) This gives hunters a power over living beings (both animals and humans) which does not arise out of their own productive work. They can not only appropriate fruits and plants (like the gatherers) but also animals and other producers (female and male) by using their arms.

(c) The object relation (mediated through arms) is basically a predatory or exploitative one.

(d) The relationship to nature (mediated through arms) constitutes a relationship of dominance and not cooperation between the hunter/warrior and nature. This dominance has become an integral element of all production relations which men have established. This is the main paradigm of their productivity.

(e) Appropriation of natural substances (Marx) slowly becomes a process of one-sided appropriation in the sense of establishing property relations and the exploitation of nature.

(f) By using arms, hunters could not just hunt animals – they could also raid communities of other subsistence producers – kidnapping their unarmed young and female workers and appropriating them. It can be assumed that the first form of private property was not cattle or food but female slaves (Meillasoux, 1975; Bornemann, 1975, as cited by Mies et al., 1988: 82–83).

Mies asserts that this does not reveal the biological aggressiveness of males. Rather it displays the dialectic relations men can create using their weapons. She postulates that this created the possibility of establishing relations which transcended the limited hunting-gathering context.

The real dominance relationship between men and women could only be established after men had discovered their own generative capacities. This discovery went hand-in-hand with the domestication and particularly the breeding of animals as a new mode of production. She concludes that this predatory mode of production became possible due to two reasons: (a) men's monopoly over arms, and (b) observation of the reproductive behaviour of animals.

After establishing these postulates, Mies infers that martial pastoral nomads were the fathers of all dominance relations, particularly those of men over women. Studies suggest that women were kidnapped for their labour power on private plots and for sale. This was the most productive activity of warrior-hunters who were not hunters-gatherers anymore but lived in an economic system based on women's productive agricultural work. This was before the introduction of the plough. Mies argues that the exchange of women (observed by Levi-Strauss) is rooted in this kind of slavery. She cites Bonte in Meillasoux's book on pre-colonial slavery and states that this slavery was the result of and the means for unequal exchange.

Surplus Generation is not the Product of New Tools but of Usurpation by 'Civilized Men'

According to Mies, the appropriation of surplus as the basis of the development of hierarchy, sexual and class relations (as postulated by Engels) does not emerge automatically from the development of productive forces. It is created by robbing, making women and men work as slaves and by armed usurping of their products. This surplus is different from the one which a community would produce and store for emergencies or future shortages. It is a forced surplus which is preconditioned on slavery or warfare (based on the male monopoly over arms). Thus warriors and conquering nomads become the agency of a predatory mode of production/appropriation. This mode sets the paradigm of all exploitative relations between human beings and nature. It developed into different forms and modes in various histories but essentially remained the same. It transformed autonomous human producers into conditions of production for others. The development of productive forces (and thus of wealth) is based on conquest and warfare in such a society. This is the origin of patriarchal societies (Mies et al., 1988: 86). In recent past American history illustrates how they exterminated native tribes and conquered the vast tracts of land and natural resources. It is known as the 'Age of Robber Barons'.

Elise Boulding explains the instability of the matrilineal system in structural terms. She states that, "Risk-takers or not, men certainly were the 'free floating resources' of this society. As the 'extra pair of hands' in the situation, they responded to the asymmetry in responsibility by inventing new patterns of social organization" (Boulding, 1976: 148).

Capitalist Patriarchy: Man the Breadwinner, Woman the Housewife

In a violence-based paradigm, the predatory mode of appropriation and colonization is apparent in feudalism and capitalism. This is the use of structural violence (where serfs are tied down to the landlord – their bodies being his property – resulting in an inability to move away from the land). The landlord maintains his monopoly by using coercive means over external enemies and internal rebellions.

Capitalism did not eliminate the former savage forms of control over human productive capacity. It reinforced and generalized them. Apparently peaceful production relations (in terms of wage labour and capital) based on the mechanism of economic coercion were built on the expansion of the predatory mode of production. This occurred first via slave trade and later through the colonial conquest of India. The best example of the predatory appropriation of capitalist economies is slavery in plantations.

Mies avers that feudal lords are different from the capitalist class because they are aware of their dependence on nature (including women). The capitalist class conceived itself as master over nature from its very inception (ibid.: 88). By looting colonies and exploiting slaves, capitalists were able to give economic and political concessions to European workers. Capitalists allowed workers to share their social paradigm. The wage labourer's colonies were the women of his own class (bound by marriage and family laws) over whom he exercised his own coercive monopoly. Domestication of bourgeois women and their transformation into housewives (dependent on the husband's income) was the model for the sexual division of labour under capitalism.

Men's proletarianization was accompanied by the 'housewifization' of women. Mies traces women's domestication from the 14th century onwards. During this period many craftswomen were pushed out of male guilds and witch hunts were carried out. Many single women, poor peasant women and urban widows were arrested, tortured and burnt (Rowbotham, 1974; Becker-Bovenschen-Brackert, 1977; Dross, 1978; Honneger, 1979 as cited by Mies et al., 1988). Sexual autonomy was closely connected with economic autonomy. 'Professional' male doctors phased out midwives and healers. Women internalised this 'civilizing process' as voluntariness and love (an ideological mystification of their own self repression).

Mies supports Wallerstein's postulate that capitalism is a world system built on a hierarchical structure of exploitation, "those who breed manpower sustain those who grow food who sustain those who grow other raw materials who sustain those involved in industrial production" (Wallerstein, 1974: 86). Mies points out that "Wallerstein forgets to mention people who sustain the non-producers (who control this whole process) by means of arms"(as cited by Mies et al., 1988: 92).

Mies stresses that 'housewife' is a necessary ideological construction of women under capitalism, irrespective of whether they actually live as housewives or not. The housewife is a model of modern woman promoted by capitalism. It is used as a strategy to internationalize women's labour in the 'informal sectors' of southern economies. After the collapse of the socialist bloc in Eastern Europe, efforts to introduce market economy were accompanied by patriarchal policies (with regard to women). In East Germany, the combined processes of colonization and housewifization were quite evident. The current policies, pursued with regard to Third World women, have been introduced to solve the crises of capitalism in the First World.

Under capitalism, women are conceptualized as housewives and men as wage labourers (who sell their labour in the market). Marx believed that this form of labour provides a base to organize workers. Organized workers can fight for liberation and establish a socialist society. Claudia Von Werlhof questions whether wage labour is more liberatory in an evolutionary order of the world (barbarism followed by feudalism and wage labour being a part of the current order of capitalism). She sees it as a process of alienation from nature. Marxist feminists accept the liberatory aspect of wage labour in contrast to slavery and bonded labour. Werlhof's paper appeared in a book called *Women, the Last Colony* which was edited by Maria Mies, Veronika Bennholdt-Thomsen and Claudia Von Werlhof (Mies et al., 1988). This book investigates how capitalism colonizes people, women and non-human nature. It declares that the world view which is built through this interaction is distorted and necrophellous.

Werlhof studies the consequences of the commodification of labour power. Humans are alienated from nature when their labour is objectified. Technological development leads to a feeling of abundance in the short term. The same technological development aspires to get rid of the dependence on women for biological reproduction. This kind of progress distorts human relationships with one another and with nature (Werlhof et al., 1988: 111).

Capitalism, Technology and Mechanical Worldview: Changing Resource Use

Mies credits Carolyn Merchant with tracing the history of the mechanistic view of the world which emerged with the industrial revolution. Merchant provides detailed description of large-scale geographical changes, which altered the face of the earth from the 17th century onwards. She also describes the changes taking place in the ideology which conceptualizes nature. Nature was previously viewed as a living organism. Central to this organic theory was the identification of nature (especially the earth) with a nurturing mother. An opposite female image of the wild and uncontrollable aspects of nature (viz., storms, droughts and general chaos) was also prevalent. The first image was replaced with the new dominant image when the scientific revolution mechanized and rationalized the world view. The second image of nature as disordered and hostile fitted with the idea of mastery.

Mechanization and domination are the core concepts of the modern world. The new mechanical philosophy of the mid-17th century achieved a reunification of cosmos, society and the self in terms of a new metaphor – the machine. Society needed these images to advocate commercialization and industrialization. Both of these depend on activities which directly alter the earth, such as mining, drainage, deforestation etc.

In a mechanical world, order implies each part behaving predictably (in tune with a system of laws) within the larger whole. Power is derived through the active and immediate intervention of a secularized world. Order and power together constitute control. Redefining reality through the image of the machine legitimizes rational control over nature, society and self (Merchant, 1983: 281).

Vandana Shiva criticizes modern science because its framework is based on a mechanistic world view. It pervades modern life and distorts our vision by attacking the integrity of nature including human beings. Modern science is reductionist and encourages fragmentation. Reductionism reduces human capacity to know nature by excluding other ways of knowing. It treats nature as inert and fragmented matter. Nature's capacity for creative regeneration and renewal is ignored. The ontological and epistemological assumptions of reductionism are based on uniformity. It perceives all systems as comprising of the same basic constituents which are discrete and atomistic.

All basic processes are assumed to be mechanical. In contrast to organic metaphors (in which concepts of order and power are based on interdependence and reciprocity), the metaphor of nature as a machine is based on divisibility and manipulability (Shiva, 1993: 22–35).

Shiva challenges the claim of modern science as being free from values. Arbitrary boundaries between knowledge and ignorance are parallelled by arbitrary boundaries between value and non-value. The reductionist, mechanistic metaphor simultaneously creates a measure of value and the instruments for the annihilation of that which it considers as non-value. It creates the possibility of colonizing and controlling that which is free and self-generative. In this sense, the seed and women's bodies are seen as sites of regenerative power. In the eyes of capitalist patriarchy, these are among the last colonies. She analyses the violence of biotechnologies and reproductive technologies in this light (ibid.).

Compatible Analytical Framework

Ecological Transformation: Mode of Production and Mode of Resource Use

In the Indian context, Madhav Gadgil and Ramchandra Guha trace a similar transition from settled cultivation to an industrial mode of production (Gadgil and Guha, 1993: 11–65, 71–90, 113–145). They theorize the transition from one production mode to another with the added dimension of ecological transformation. This provides an analytical framework for the expositions of the three feminists presented above. However, this framework ignores women's role in production, reproduction and in the belief system. Gadgil and Guha seem to believe that culture and ideology are men's prerogatives.

Gadgil and Guha attempt to understand the characteristics of any society using the mode of production concept, proposed by Marx. This concept constitutes economic activities which provide the infrastructural base. The Marxist theory proposes that economic activities are shaped by the antagonistic interaction of two categories of the mode of production: (a) relations of production, and (b) productive forces. When the relations of production become a fetter to the development of productive forces, society ruptures and a new mode of production emerges.

Gadgil and Guha claim that the mode of production concept is not adequately materialistic as it does not take into account the ecological context (i.e., soil, water, animals, minerals and vegetation). Society's economic infrastructure is embedded within this context. Marx treated nature as an abundant source of raw materials (i.e., natural resources). Thus, the man's relationship with nature was not included in Marx's inquiry. Gadgil and Guha want to complement the 'mode of production' concept with 'modes of resource use' concept. The latter extends the realm of production to include flora, fauna, water and minerals. With respects to the relations of production, it investigates the forms of property, management and control and of allocation and distribution. These govern the utilization of natural resources in different societies and historical periods.

With respect to productive force, this perspective analyses the technologies of resource exploitation, conversion and transportation (which characterize different social orders). The scheme of resource use incorporates two additional dimensions within the mode of production framework. First, it examines whether one can identify characteristic ideologies which govern different modes. More importantly, it identifies the ecological impact of various modes thereby assessing their consequences on the pattern, distribution and availability of natural resources.

Gadgil and Guha broadly compare four modes of production across five different axes. The four modes of production are defined as gathering, nomadic pastoralism, settled cultivation and industry. The five axes are defined in the following manner:

- *Technology:* energy sources, materials used and the knowledge base related to resource use.
- *Economy:* the spatial scale of resource flows and modes of resource acquisition.

- *Social organization:* size of the social group, the division of labour and mechanisms of control over access to resources.
- *Ideology:* broad perceptions of man-nature relationship as well as specific practices of promoting resource conservation or destruction.
- *Ecological impact:* impact on nature and its regenerative capacities.

The Industrial Mode of Production and Modes of Resource Use

Technology

As a mode of resource use, large-scale industries are barely two hundred years old. This is only one-fiftieth of the time which Homo sapiens have spent husbanding plants and animals and only two hundredth of the time since when hunter-gatherers painted hunting scenes on cave walls. However, their economic impact has surpassed all preceding modes. This is because of a quantum jump in energy use (with heavy demands on non-renewable sources coupled with novel sources such as nuclear energy). Resource flow has become global due to material processing, storage and transportation facilities. A small fraction of humanity has come to consume vast quantities of a wide range of resources.

Economy

Industrial societies consume resources at an accelerated rate thereby exhausting them sequentially – from economically desirable resources to less desirable ones. European colonists usurped vast tracts of land around the world because of the energy and transport crisis brought about by the exhaustion of their forests. Gadgil and Guha reconstruct the history of colonization as a history of unsustainable resource utilization. This perpetuated a need for conquest to maintain the same mode of resource use. Such resource utilization is characterized by asymmetric resource flows. Industrial societies import large volumes of cheap unprocessed resources while exporting smaller volumes of unprocessed resources at much higher prices. This process produces inordinately high amounts of waste in industrial societies. These are disposed either in the global commons (oceans and atmosphere) or sold to Third World societies plagued by foreign debt.

Social Organization

The social organization of the industrial mode of production is marked by the rise of individualism. Ironically, this is accompanied by a tremendous expansion of the state's role in regulating individual transactions. For instance, the modern industrial state safeguards private property (land, workplace etc.) and assumes ownership over what was hitherto common property. In the process, it de-legitimizes community-based systems of access and control.

Ideology

Industrial society rejects the gatherer (part of a community of beings) and the agriculturist (a steward of nature) view of man. It emphatically asserts that man is separate from nature and has every right to exploit natural resources for his own well-being. Hence nature is de-sacrilised and the market place is venerated. The market is supposed to rationally locate the efficient use of resources so that all individuals are as well off as they could possibly be.

Gadgil and Guha are concerned about the claim that scientific enquiry into nature (over the last century) has made it possible to develop prescriptions on the sustainable use of soils, forests, fisheries and other renewable resources. Religion and customs legitimized patterns of resource use in pre-industrial societies. In the industrial mode, science is the organizing principle for human interactions with nature. Scientific prescriptions for the preservation, sustainable use and the deliberate destruction of different species and habitats parallel pre-scientific prescriptions typical of earlier modes. These were based on an informal knowledge base or on simple rule of thumb. Modern prescriptions are based, at least in part, on a more detailed knowledge of the system and on a more explicit definition of what is sought to be achieved.

Gadgil and Guha challenge the superiority of these scientific prescriptions which claim to achieve maximum sustainable yield (i.e., a yield which will not decline in the long run). Every plant or animal population is involved in a number of interactions with other members of the community. Their argument is that because of the complexity of these interactions, the simultaneous harvesting of more than one resource population (from a specific locality) may have unexpected consequences. For instance, scientific prescriptions for maximum sustainable yield in the artificial plantations of Europe have largely failed except in very limited contexts. Even in these contexts, the gradual exhaustion of soil nutrients is now leading to a decline in the forests.

Ecological Impact

Spatially, hunter-gatherers live in the forest; agriculturists live adjacent to but within striking distance of the forest and urban-industrial men live away from the forest. Paradoxically, the greatest impact on forest ecology comes from people who are the most spatially separated from it. Hence, they are also further removed from the consequences of this impact. The same conditions operate with regard to other resources (like water) as well.

While admitting the importance of scientific knowledge, the authors assert that the idiom of resource use is more important. Modern science has enlarged our knowledge of specific physical and chemical processes. However, it does not always display a similar understanding of the ecological consequences of human intervention. The belief that science provides an infallible guide has nonetheless encouraged major interventions in natural ecosystems resulting in unanticipated and usually unfortunate consequences. The history of fisheries and forest management are replete with illustrations of the failure of sustained-yield methods in forestalling ecological collapse.

Gadgil and Guha point out that contemporary environmental movements express a widespread disaffection with the excesses of the industrial mode. It is no mere coincidence that methods drawn from religion and custom (as opposed to 'modern' scientific ones) are upheld as being more prudent in their use of nature.

Gadgil and Guha present five distinct but interrelated processes of the industrial mode: (a) an increasing intensity of resource use and exploitation, (b) a secular increase in resource flows across different geographical regions and across different levels of any economic/political system, (c) an integration of larger areas into the domain of any given political/economic system, (d) a secular increase, at the global level, in population densities and in the extent of stratification and inequality with respect to the access, control and use of different natural resources, (e) finally, an intensification of the rate of ecological change and ecological disturbances

By mapping the larger processes which connect people with nature and by understanding the dynamics of institutional structures (such as whether the gatherer, feudal or capitalist mode of production is being followed) along with the modes of resource use (technology, natural resource management systems etc.), we can gain a wider understanding of our reality. It is useful to chalk out a strategy against these ecologically-destructive processes.

Gadgil and Guha have not considered resource use by the sex/gender system. Thus, they do not segregate the knowledge base and the resource use pattern by gender. A segregated analysis will help in designing a system of sustainable resource management and appropriate institution-building processes with the people. There has been an effort to create access to natural resources for the resource poor through newly created institutions. In Joint Forest Management (JFM) Committees and Watershed Committees, women, particularly single, landless or commons-dependant women, are not given sufficient representation. Considering the family as a unit for entitlement ignores the special needs of these women. JFM committees mainly discuss the issue of sharing timber which earns cash for the families. This issue is of greater interest to men. However, there is no discussion about who will be allowed to collect firewood and in what quantity. Generally, it is the poor (possibly younger women including daughter-in-laws or girls) who carry out these tasks. However, they are not represented in these committees leading to a further deterioration of their condition. This kind of exclusion affects the efficiency of these management systems. Women collect firewood illegally and sell it in the market without imbibing a sense of ownership over the forest. Taking into account the nature of subsistence work and women's lack of mobility, they should be given an opportunity to take leadership roles in the development of new resources. In development parlance, they are the primary stakeholders.

6

Symbolic Representations of Nature and Women

West and India

> American Indian story of the genesis of the agro-ecological system is based on the abundance of interdependent poly-cultures in forest gardens. Nature is a desert, poor place for human existence and woman is perceived as the saviour. Through a willing sacrifice in which her body is returned to the earth, she introduces agriculture to her husband and to women who subsequently plant the corn, beans and squash that provide the bulk of the food sustaining the life of the tribe. The earth is treated as an agent of regeneration. Death is transformed into life through reunification of the corn mother body with the earth. Even death results in a higher good. (Merchant, 1995)

Ecofeminists claim that the identification of women with nature has been expressed throughout history. It permeates the philosophical basis of scientific methodology. A strand of cultural ecofeminism believes that Mother Goddess symbolisms and the period in which they were created highlight the fact that women are not necessarily crippled by caring and reproductive responsibilities. Cultural feminists emphasize that these functions are trivialized and posited as barriers to women's empowerment under certain patriarchal institutions. Hence, they seek to discover old rituals. They claim that these symbols signify a new awakening of holistic life practices. Reproductive responsibilities and caring assume centre stage and are shared by men within such practices.

In contrast, Marxist and liberal feminists do not analyse the pre-historic period of Mother Goddesses as they find it purely conjectural. They claim that men have used this identification to belittle women and nature. It has denied agency to women throughout patriarchal history. Under the same regime, women's subordination is reinforced through myths and social practices. Women striving for equality and emancipation in public life see this identification as a hurdle. According to them, ecofeminism propagates essentialism by concentrating on the biological aspect of women's lives.

Two authors are cited here who have analysed the symbols of Mother Goddesses in Europe. Riane Eisler, in her book *Chalice and Blade* (1987) interprets the materials found in some newly excavated archaeological sites in Greece and Eastern Europe. Carolyn Merchant traces the representation of goddesses in European and American religious literature and cultural productions during the Middle Ages and later during industrialization's concurrent expansion of cultivation in America.

In India, there is ample literature (primarily ancient scriptures) depicting nature in terms of a feminine principle. These symbols are rarely invoked by Indian ecofeminists. However, there is an urgent need to historically trace the changing imagery of nature and linking it to technological and economic changes. Maithreyi Krishnaraj edited a book called *Motherhood in India: Glorification without Empowerment?* (2009), in which the investigative work of feminist cultural anthropologists on the archaeological findings of temples and icons is presented. These anthropologists link their findings to rituals of the 'little tradition' and the 'great tradition'. They attempt to study the status of women in the pre-history period. There is a striking similarity in the images found in European archaeological sites and the world view propagated through Indian imagery.

Western History

Nature as Mother Earth

The history of almost all cultures is replete with the identification of nature with women. Similarly, all ancient religious literature regards the act of regeneration as sacred and divine. The image of the earth as a mother and a living organism (Gaia) has been constructed ever since early civilizations began.

The Mother Goddess cult in India and Egypt is well known. Recent archaeological excavations (at Catal Huyuk near Greece and near Poland and Hungary) have revealed primary information about the Neolithic period. These are significant because Europe was considered the epicentre of pastoral tribes and hence less likely to support matrilineal communities. By re-evaluating and reconstructing the past, Riane Eisler (1995: 369–379) throws light on this pre-historical period. Her analysis is based on information obtained through excavations of buildings and their contents including clothing, jewellery, food, furniture, containers, tools, articles at burial sites etc.

Eisler observes that the art found in these excavations neither displays any emblem of power nor any imagery idealizing armed might, cruelty and violence. In contrast to male civilizations, there are no lavish burials of 'chieftains'. The murals, statutes and votive figurines are images of goddesses. In her various incarnations of Maiden, Ancestress or Creatrix, she is the lady of the water, birds and the underworld or simply the divine mother cradling her divine child in her arms. Some portrayals are surrealistic or grotesque in form which may be an attempt to deal with the darker aspects of reality (such as human fears). She insists, "that with all this, the many images of the Goddess in her dual aspect of life and death seem to express a view of the world in which the primary purpose of art and of life, was not to conquer, pillage and loot but to cultivate the earth and provide the material and spiritual wherewithal for satisfying life" (Eisler, 1995: 372).

Eisler notes that goddess worship was both polytheistic and monotheistic. It was polytheistic in the sense that women were worshipped under different names and in different forms. But it was monotheistic in the sense that faith in the Goddess was the same as the contemporary faith in God as a transcending unity. Agricultural societies which worship the Goddess also deified the female who gave birth and nourished (just like the earth).

Eisler objects to scholars who refer to goddess worship as a 'fertility cult'. She feels that the people who worshipped the Goddess were deeply religious because at that time there was no separation between the secular and the sacred. Religion and life were inextricably linked. Religions in which the most powerful or only deity is male tend to reflect a social order in which descent is patrilineal and domicile is patrilocal. The opposite is true for religions with predominantly female deities. She insists that the archaeological findings at Catal Huyuk do not reveal a matriarchal society. Despite women's pre-eminence in religion and life, there is no indication of any inequality between women and men. Nor is there any proof that women subjugated or oppressed men. Thus, Eisler posits that goddess worshipping societies were ecological and egalitarian societies. There seems to be a close connection between women's status and the status of nature (i.e., ecological balance). Reverence for regenerative powers appears to be a common indicator.

Eisler's observation sounds true in the case of monotheistic religions like Christianity and Islam. However, it appears faulty in countries like India where goddess worship co-exists with the worship of male gods and their consorts. The goddesses are portrayed as autonomous. We also have a lot of *gram-devis* which are considered 'living' goddesses in a hierarchical class society (Ganesh, 2009).

Changing Representation of Women

Gaia

Carolyn Merchant traces the reversal of the supremacy of women and nature (during the ascent of dominators' society) through her deconstruction of the narratives of three goddesses: Gaia (the Greek Goddess), Eve (the Biblical Goddess) and Isis (the Egyptian Goddess) (Merchant, 1995: 3–74). She views the transformation of these narratives along with the socio-economic changes in society. The changing representation of women is the result of the changing needs of dominant patriarchal forces.

Merchant selects Gaia (also called Ge) or the ancient Mother Earth who was studied by Charlene Spretnak in her paper "Lost Goddesses of Early Greece" (1978). Spretnak's goal was to reclaim pre-Hellenic goddess myths (which existed prior to the transformation of Greece by barbarian invaders) and to restore them to forms predating their patriarchal codification by Hesiod and Homer in the 7th century. Gaia was the ancient Mother Earth who brought forth the world and the human race from "the gaping void, chaos".

Spretnak's identification of Gaia as a powerful feminist and ecological symbol was followed by Lovelock, who scientifically proved his concept of homeostasis in 1979 (without acknowledging that the basis of the Gaia hypothesis was provided by Spretnak). Lovelock states that homeostasis (restoration of equilibrium) is the process by which nature restores its balance through inherent regenerative processes over time. He agrees that the concept of Mother Earth (or Gaia of the Greeks) has been the basis of a widely held belief system throughout history which still co-exists with the great religions of the world. Merchant mentions that feminists were disturbed by the term 'homeostasis' as it assured that nature would eventually regenerate itself despite any kind of problem created by humans. Feminist criticism points out that the concept reflected the typically male attitude of women cleaning and tidying up their mess. Feminists were indignant that men never wanted to take responsibility for their actions.

Merchant narrates that with the introduction of the Gaia concept the ecofeminist discourse started strongly emphasizing the connections between women and nature. The term 'ecofeminism' was coined by the French writer Francoise d'Eaubonne in 1974. Ynestra King is credited with popularizing the concept. The entire movement in 1980 rallied around this term after the conference on "Women and Life on Earth" in Amharst, Massachusetts.

Many strands of ecofeminism have emerged since then. However, only cultural ecofeminism celebrates the revival of ancient rituals centred on goddess worship, the moon, animals and the female reproductive system. Spirituality is seen as a catalyst for personal and social change. Women's biology and nature are celebrated as the sources of female power. Cultural feminists inverted the radical feminist view of the connection between women and biological reproduction being a barrier to women's creation of culture. They asserted that women have their own culture which has thus far remained invisible. This culture is different from the current domineering patriarchal culture. It has become the source of women's empowerment and ecological activism.

Eve

The second goddess deconstructed by Merchant is Eve. The story of Eve is narrated through different Biblical versions and art forms (including those used in America by European colonizers to justify their settlement). Merchant analyses how the connection between images of woman and nature is constructed to suit capitalism. She contrasts the American Indian story of the genesis of the human civilization (which reflects their holistic understanding of nature and their symbiotic relationship with it) with the American/European story. The latter is a story of men dominating women and nature. The American Indian story focuses on the willing sacrifice of the corn mother. She teaches agriculture to her husband and the entire community by returning to the earth. Hence, the earth becomes an agent of regeneration.

Puritan Christian fathers brought their own agricultural origin story into this bountiful world of corn mothers. The Biblical story begins where the Indian story ends. A woman, Eve, shows the man, Adam, how to pick fruit from the tree of Knowledge of Good and Evil. Instead of attaining good, the couple is cast out of the garden into a desert. Merchant points out that the couple moves from desert to garden in the Indian story while in the Biblical story from garden to desert. The fall from paradise is caused by a woman and a man becomes the agent of transformation. This is a plot of fall and recovery. European settlers used the same theme in their creative endeavours. They also displayed a passionate desire to change the earth (in this case the wilderness in America). Modern Europeans added mechanistic science and laissez-faire capitalism to the Christian recovery project to create the master narrative of Enlightenment.

Merchant notes that a crucial part of the plot structure is conceptualizing nature as female. The rhetoric of American settlement is replete with language which casts nature as a female object to be transformed. Men are viewed as the agents of change. Female images of nature are encoded into the texts of American history, art and literature. They function as ideologies for settlement.

In the recovery story, Eve is virginal, pure and light. The land is either pristine or barren but with a potential for development. Like the fallen Eve, nature is disorderly, chaotic, a wilderness or wasteland, a desert requiring improvement, dark, witch-like and the mouthpiece of Satan the serpent. Later, (in the 19th century), male agriculturists used the plough technology to compel female nature to produce. The narrative of frontier expansion is the story of male energy subduing female nature, taming the wild, ploughing the land and recreating the garden lost by Eve. Merchant cites the imagery used by male authors to describe the seduction of female earth (which implies violent rape). Civilization was considered the ultimate end in which the unruly energy of wild female nature was suppressed and pacified. The final natural state is depicted as female and civilized – the restored garden of the world.

Isis

Merchant de-constructs the history of science through Isis. George Sarton, the founding father of the history of science, used the name of the Egyptian Mother Goddess Isis to represent nature. For Sarton, Isis's robe concealed nature's secrets. The goal of positivist science was to solve nature's mysteries by disclosing the secrets that 'she' harboured within. The image of science as positive knowledge, its construction out of atomistic textual nuggets and the methodology of "wrestling with nature" to decode 'her' secrets are deeply imbedded in the history of

Western culture. These assumptions about science, nature and method pervade both scientific and historical texts. Aristotle identified the female principle with passivity. He further associated passivity with the object and the natural world. These assumptions furnish the basic philosophical framework of Western culture which maintains a hierarchical domination of subject over object, male over female and culture over nature.

Scientific positivism was accentuated when science chose to depend on visual textual material and the study of objects through scientific instruments. Western culture's penchant for ocularcentrism (i.e., communication through visual representation) imbues these images with power. Representation is a sign or a narrative account intended to influence, persuade or be interpreted by a viewer or hearer. Science makes people believe that it knows nature by representing nature through written texts, illustrations, metaphors and symbols. Many writings by science historians express the fervent desire of being able to lift the robe of Isis (i.e., the veil of nature). This means that matter itself is to be decoded to allow us to view the hidden secrets of the atom.

Merchant exposed the gender and race biases of the historians of science. She shows that Sarton was eager to establish the pre-Hellenic roots of science among Egyptians, Babylonians and Hindus because he saw them as precursors to the "Greek Miracles". Merchant mentions the debate about the black roots of Egyptian science. Images of Isis frequently depict her with black or brown skin and curly hair. Merchant notes that in the 17th century, matter was associated with black magic, witchcraft and black African slaves. Hence, a need was felt to achieve rational, scientific and technological control over nature to prove the superiority of the white race. Isis was slowly appropriated as a white Goddess. This appropriation is expressed in Athenasius Kircher's depiction of Isis with European features (including lightened skin and hair).

Merchant quotes Londa Schiebinger who has investigated the theories of gender and race at the root of modern science. She claims that by the 18th century, black women's bodies, skulls, breasts and genitalia had become objects of investigation. This was defended on the grounds of a broad scientific inquiry into the anatomical and sexual differences between Africans and Europeans.

Sarton's account of the progress of science is replete with men setting the standard for knowing and culture under the belief that they fulfil "the destiny of the human race". By dealing with the representation of nature (specifically the treatment of goddesses), Merchant exposes the processes which dominate nature and women. She provides the basis for the cultural feminist assertion that women's subjugation is not based on biological reality.

Postmodernist feminists and the Frankfurt School[10] expose the universalizing tendencies prevalent in Western culture since the Enlightenment (in the 18th century). They criticize the idea that history is linear and evolutionary and that reason reveals transcendent truths. They question whether reason can be separated from the body and from particular times and places. They also challenge the authority of science to make universal claims about the body, particularly female bodies.

Ancient Indian Literature and Symbolic Representation of Woman and Nature

There has been no systematic attempt to historicize the connection between images of women and nature in Indian literature thus far. Some superficial attempts at connecting ancient philosophies, belief systems and prevalent customs are found in environmental literature. In *Staying Alive* (1989), Vandana Shiva mentions that ancient Indian literature (particularly Sankhya philosophy) identified women with nature as both had regenerative powers. She uses this to illustrate that pre-industrial societies lived in harmony with nature.

Gadgil and Guha (1993) have studied the customs and belief systems which used to limit resource use thereby enabling people to maintain a balance between exploitation and regeneration. However, their analysis is completely gender blind. It glosses over the bifurcation of resource use in the industrial mode of production. They do not theorize the patriarchal system which controls women's sexuality to perpetuate property ownership in the hands of men and property owning castes. Women are granted access only to the commons for fuel and drinking water. The nurturance and regeneration of these commons is their responsibility.

Neither Shiva nor Gadgil and Guha have studied this historical period and noted down women's role in this mode of production. Shiva is often criticized of glorifying women's role in resource management in pre-industrial societies. Invoking representations of women in the distant past may not be relevant to contemporary times where women are marginalized (i.e., during the transformation of agriculture from subsistence to cash crop). Gadgil and Guha present the process of transformation at a very abstract level. They do not attempt to understand women's roles even in the belief system.

Shiva's invocation of women's role in the nurturance and regeneration of nature cannot be completely brushed aside as being outmoded. This is especially true in a country like India where more than half the cultivable land is under subsistence production and is cultivated mostly by women. Krishnaraj observes that women in agriculture are currently operating in a transitional economy in which pre-capitalistic and domestic modes co-exist with an increasingly monetising commercial, market economy. Unlike in the West, domestic technology has remained unchanged. There are not enough market substitutes to diminish household provisioning. The number of family households supported entirely by women has increased by more than 30–40 per cent because of male migration (Krishnaraj, 2008: 30). Wood fuel comprises 80 per cent of the cooking medium and is collected by women. They are also responsible for procuring drinking water for the household. Hence women are concerned about preserving the primary productivity and the regenerative capacity of natural resources. Even though they have not had a say in resource management within institutional forms, they have expressed concerns for sustainability more often than their menfolk.

There are three philosophical discourses which use feminine representation (mainly the imagery of Mother Goddesses) to depict nature. The first strand is found in pre-Vedic traditions (Neolithic period), images of Mother Goddess and in belief systems which have survived in folk literature and customs. The second discourse is Sankhya philosophy which is invoked by Vandana Shiva and is known to have materialistic roots. The third is the Vedic discourse recognized by high-caste Hindus as the source of the Hindu tradition.

Pre-Veda Period, Mother Goddesses and "Fertility Rites"

Debiprasad Chattopadhyaya (1959)[11] and D.D. Kosambi[12] (2009) have documented the Mother Goddess tradition. They have studied folk beliefs and customs which originated in pre-Aryan India (mainly in foraging-gathering and horticulturist societies). In his paper, "A Study of Mother Goddess Cult Sites" Kosambi (2009: 830–859) has written about small tablets of goddesses (as protective *gram-devata*) found in the Deccan plateau of Maharashtra.

S.R. Gadgil (1998) has collected many anthropological references in his Marathi book *Lokayat*. He cites many sayings by Khasi tribes including "from the woman sprang the clan", i.e., the women started settled societies. By citing examples of Mother Goddess, he confirms Briffault's analysis that primitive societies followed the matriarchal system (today, these societies are called matrilineal and not matriarchal). Men were mobile and made weapons for hunting. Women were stable and hence invented horticulture. They built storage facilities (straw baskets, mud pots etc.) for grain. He also cites many land fertility rites, such as spreading menstrual blood before cultivation or the performance of mass copulation before ploughing. Women were responsible for invoking rain gods. These rites involved women who had been proven fertile – a mother who could invoke the fertility potential of land and help her regenerate and produce.

R.C. Dhere in his book *Lajjagouri* (1978) (one of the many names given to Mother Goddess) has collected photographs of Mother Goddess icons. These are not worshipped any longer. Sometimes they are relegated to a corner of the temple. They may even be buried under a new temple of the male god Shiva. These statues are mainly found in South Indian states including Maharashtra. Gauri seems to be the generic name of all Mother Goddesses associated with fertility rites. He formulates that after the ascent of Vedic culture, these goddesses were converted to consorts of male gods. The new gods were established in temples after removing older idols. The Devdasi tradition seems to have its roots in this period when women were allowed to act as priests. Through patriarchal ascent they were relegated to the status of concubines (and later as prostitutes in brothels). It is essential to connect these to the changing modes of production, resource use and resource flows.

Krishnaraj, in her introduction to the book *Motherhood in India* maps the literature on the goddess worship practices of Manoshi Mitra, Giti Thadani and others. Giti Thadani has found abundant evidence of an earlier celebration of feminine principles prominent in cultural details depicting the generative power of the mother through explicit depiction of the 'yoni', (which was considered a source of energy). Krishnaraj cites Thadani's observation that many temples became masculinized when independent female goddesses were converted into the consorts of male gods. These old temples did not have a sanctum sanctorum. Rather they had an amphitheatre which was supported by pillars and was open on all sides (suggesting the decentralization of energy) (Krishnaraj, 2009: 17–18). Kamala Ganesh's paper in the same book reminds us that the concept of the goddess has its philosophical roots in Samkhya philosophy (with its categories of Purusha and Prakriti).

Samkhya Philosophy: Does Prakrti, a Feminine Principle, Play a Central Role?

Samkhya philosophy provides a materialist explanation of the origin of the cosmos. The cosmos is derived out of an intercourse between the female principle (which represents matter) and the male principle (which represents spirit). According to this philosophy, the cosmos is a living organism. Samkhya philosophy endorses many fertility rites which were a part of early societies engaged in subsistence farming. This is in contrast to the metaphysical philosophy of the Vedic period (propagated by pastoral Aryan tribes) in which God is male. The Aryan tribes are assumed to have arrived in India from the North. They conquered native tribes which were organized on the matrilineal system and used the subsistence farming system.

According to S.R. Gadgil (1998), the male principle in Samkhya philosophy is envisioned as abstract, sensuousless and without dynamism. The female principle is full of life and begets life. She experiences the world (human and non-human) around her through all the senses and identifies with it. It appears that later Sankhya disciples were influenced by Vedic philosophy and transformed the role of the male principle (which acted as the spirit behind the matter – the knower of the world, the conscience). The female principle wooed this male into worldly experiences. She was supposed to have created the entire living, sensuous world to please this male principle.

Gadgil writes that Samkhya philosophy was later adopted by the Tantrik cult which professed to be the most materialistic religion. It believed that even men have centres in which the female principle is located. These sensuous centres are mapped on the body using the image of a triangle which represents the vulva. To invoke sensuous experience through these centres was supposed to be the act of becoming one with the female principle, i.e., with the universe. Every disciple of the Tantrik cult wanted to feel like a woman. A ritualistic intercourse considered to enable one to achieve salvation and the highest level of happiness.

However, the explanations of Samkhya philosophy vary among scholars. According to M.P. Rege (1995: 115–17), the Samkhya theory divides all things into two radically different groups: (a) things which possess consciousness or rather are constituted of consciousness (chetana), and (b) things which are unconscious (jada) but are the objects of (or for) consciousness. For Samkhyas, physical objects, organisms and minds are on one side of the ontological divide, while the other side is constituted by subjects of pure consciousness (Purusa). The totality of jada things is called Prakriti. Prakriti is one in two ways: the formal sense of being the all-inclusive set of jada things and in being one causal system. Prakriti is a feminine word and principle.

Rege explains that the ultimate constituents of Prakrti (or all jada) are called gunas. Guna is a substantive entity. It does not possess any quality. Rather guna is an entity with its own characteristic expression. Guna falls into three kinds – Sattava, Rajas and Tamas. For Sankhya, a jada thing is something which is essentially a (potential) object of consciousness, an object for Purusa. The consciousness of Purusa is likened to light. For Purusa to be aware of an object is for the object to be illumined with a glow from the light of Purusa. Sattva guna enables the jada to glow in the light of Purusa.

Rege describes that only the preponderance of Sattva guna allows this mental phenomenon. This is a conscious experience. Opposed to Sattva guna is Tamas guna which resists the light of Purusa. This guna constitutes the principle of materiality. The third kind of guna (the Rajas guna) expresses itself in motion. It constitutes the dynamic principle in Prakrti. Tamas guna, in addition to resisting the light of Purusa, resists the force exerted by Rajas guna.

Every phenomenon in Prakrti is an admixture of all the three kinds of gunas. The proportion in which they are mixed determines their character. Things cannot be separated from one another by assigning them to different species or sorts (each with a sharp boundary). Rather they form a continuous gradation stretching from the most Tamasika things to the most Sattvika.

Rege compares Samkhya dualism with Cartesian dualism which dominated Western thought during the Enlightenment. According to Cartesian dualism, the basic ontological division is between the mind (characterized by consciousness) and material things (characterized by extension and motion).

Rege's explanation of the Samkhya theory does not interpret Prakrti as a feminine principle and Purusa as a masculine principle. Although the concepts of Prakrti and Purusa appear as a dualist pair (comparable to Cartesian logic), Sankhya philosophy treats matter as a continuous reality without sharp boundaries between species or different things. Hence, this philosophy can be the basis of holistic conceptualizations which are valued in ecological science. Rege's presentation has a patriarchal bias as it highlights the Purusa element. He is criticized of introducing a Hinduized version of the Samkhya theory.

Vedic Literature: Goddess Ecology

Many writers recognize that a feminine conceptualization of nature occupies a significant place in Indian religious history. Almost all geographical features of the natural environment are personified as goddesses. Mountains, caves, rocks, forests, trees, plants, healing herbs, rivers, streams and lakes were conceived as potent symbol of feminine power inherent in nature. Madhu Khanna notes that, "life sustaining foods are also personified as goddesses such as Annapurna, Sataksi and Sakambari. Thus, from the sky wandering celestial bodies to the sprouting plants, all were conceived of as a manifestation of the feminine principle. In later literature these personifications culminate to form an all inclusive cosmic form (virat swarupa) of the goddess where mountains, rivers, celestial bodies, vegetation and the stratum of space form various parts of her body" (Khanna, 1995: 103–116).

Khanna avers that all the nature personifications of the goddess invariably converge into the abstract notion of Sakti or energy and its inherent power of synergy. The hidden feminine Sakti (inherent in nature) stirs seeds to fructify as the universe. Sakti is the energetic feminine potency of the Mother Earth. The earth is a live, animate organism with an efficient network linked to a biological vision of the ecosystem.

Khanna cites Brahadaranyaka Upnishad (2.1.19) which employs the metaphor of a spider sitting at the centre of its web, issuing and re-absorbing its threads in a concentric circle. The spider's web symmetrically expands to a visible circumference. Though there are divergent lines, all threads interrelate and are traceable to the central point. The sacred geography of land is a reflection of this interrelated web. Earth, water, plants, animals and human life are interdependent and interrelated in the grand design of creation. The fundamental motive behind this ancient perception is the belief that nature is conscious, alive and animated. Khanna attempts to undertake a comprehensive analysis of the conceptual understanding of the feminine principle as an integral symbol of nature. She specifically looks at the earth and waters of life, i.e., the descent of the Ganges in the light of ecological concerns.

Khanna claims that the descriptive, prescriptive and ecological passages on nature deities in scriptural sources were a means to inculcate ecological awareness. She has explained the concept of Mother Earth as a supporter in the primal experience of being an offspring of the earth. The Mother Earth is seen as the vitality which generates growth and germination in her maternal womb. The image of the earth as a universal womb explains why vessels, hollows, grottoes and caves are treated as the vulvic body of the Mother Earth. Even today, we use the 'kumbh' as a sacred vessel for rituals. It is meant to symbolize the 'womb'. The earth is sustained by the harmonious cosmic principle of Rata (i.e., cosmic order). It is a self-regulative law of harmony. According to Khanna, the Atharva Veda (12.1.60) espouses a strategy with a sensitive attitude towards the integrity and stability of the biotic community. The following hymn (where the earth is invoked with a feeling of great humility) is offered as an example:

> Whatever, I dig of you, O Earth,
> May that grow quickly upon you,
> O Pure One, may my thrust never pierce thy
> Vital points, Thy heart (Atharva Veda, 12.35). (Khanna, 1995: 111)

Indian feminist scholars are wary of claiming our legacy of Mother Goddesses. Kamala Ganesh reflects upon the relative paucity of feminist scholarly writings on goddesses. She claims that the complexity of the Indian scenario (where goddesses are a living tradition, where the abundance of ethnographic and textual materials has created a considerable non-feminist body of writing, and where there is a political mobilization of the goddess tradition) has made it a tricky cause for feminists to espouse (Ganesh, 2009: 93). However, ecofeminists like Maria Mies wants to reconnect the tradition with contemporary poor women in rural areas of India and the Third World in general. She asserts that poor women in rural areas know that the earth is a living being which guarantees their own and their fellow creatures' survival. They respect and celebrate the earth's sacredness through many rituals which are still living traditions. They also resist transforming her into dead raw material for industries. This materialism is rooted in the everyday subsistence production of the majority of the world's women. It is not an abstract essentialism for them to identify a closer relationship with nature. Their survival is linked to nature's health.

7

Epistemological Challenges

Enquiry into History of Woman-Nature Connection

> In seventeenth century England, witch mania reached its apogee. The new mechanical vision provided a secure intellectual domain for masculinity by excluding even allegorical cooperation between male and female, both in its picture of the scientist and in its picture of nature. For Baconians chastity was a precondition for Godliness and Truth whereas the emphasis placed by alchemists on the powers of love and kinship between sexual intercourse and knowledge threatened to embroil the new science and endangered science's emerging purity which taken seriously, having effective powers, provided a natural focus of anxiety. Baconians required invoking of the Devil (the witch). It indeed helped promote the simultaneous vanquishing of nature and female voracity. At that point they were not ready to consider nature as machine, which was done by Cartesian atheism a little later. (Keller, 1985: 59)

Ecofeminists argue that the woman-nature connection and men's domination over both are the cause of ecological destruction and the exploitation of women under capitalist patriarchy. Val Plumwood has traced the epistemological basis of the modern scientific world view to 'dualistic rationality'. This view is responsible for the inferiorization of women and nature. Maria Mies blames the Cartesian rubric for the subjection of women and nature to scientific enquiry. Such enquiry is carried out with the promise of equality with men and freedom from 'reproduction' (realized through the technological manipulation of nature). The developmental model built by capitalist patriarchy uses the same mechanistic, reductionist logic to manage natural resources.

A historical analysis of the construction of (so-called) scientific knowledge involves a study of the concepts of 'Cartesian rubrics' and 'dualistic rationality'. Modern civilization excludes knowledge constructed by women through their interaction with nature. Several feminists from different disciplinary backgrounds have exposed a male bias in the scientific method. This method is described as autonomous, objective and neutral. It is supposed be the basis of a higher level theory which can explain the laws of nature in a way which is better than that of experience based traditional knowledges. This critique is relevant to the politics of development which is formulated differently by different streams of feminism.

Evelyn Fox Keller (1989: 33–34) has analysed the parallels between the two systems of gender-sex and science-nature. She wonders whether gender is an analytical category different from race, class etc. She wishes to ascertain whether scientific claims to knowledge are any better than non-scientific ones. Recently feminist studies and science studies have begun to converge. This development requires gender to be seen as distinct from sex and science as distinct from nature. Keller asserts that science can never be a 'mirror of nature'.

Her statements are related to a new understanding of the 'nature of scientific method' proposed by Thomas Kuhn, an American historian and philosopher of science. In his book *The Structure of Scientific Revolution* (1962), Kuhn argues against logical positivism – a post-war philosophical movement – which emphasized on 'objectivity' and the 'neutrality' of its techniques like experimental testing (variability). These allowed scientists to compare their theory with facts. Thus, for them, science was the only paradigmatically rational activity. Logical positivism combines empiricism with a version of rationalism which incorporates mathematical and logico-linguistic constructs and epistemological deductions. Conventional scientists distinguish between theories and observational facts. They demarcate the 'context of discovery' (the actual historical process by which a scientist arrives at a given theory) from the 'context of justification' (by which a scientist tries to justify his theory once it is already there) (Okasha, 2006: 78). The ideal of theory neutrality is an illusion as data is invariably contaminated by theoretical assumptions (because of the subject-object duality). Kuhn questions whether the notion of objective truth makes any sense at all. The idea that there are fixed facts about the world (independent of a particular paradigm) is of dubious coherence. According to him, a 'scientific revolution' occurs when the scientific community undergoes a paradigm shift. This shift may not necessarily be due to the internal logic of scientific activities. It may be caused by the prevalent social factors of that time. Currently, science is a 'normal science' (which is like solving a puzzle). It does not have the potential to make earth shattering discoveries.

The period of normal science lasts many decades. As scientists finetune their theories, anomalies are discovered. These break confidence in the existing paradigm. A variety of alternatives are proposed and eventually a new paradigm gets established (ibid.: 81). Krishnaraj defined paradigm as the fundamental image of subject matter. It subsumes, defines and interrelates problems and the rules evolved. These are used to solve other situations and the methods and instruments of obtaining knowledge, analysis or solutions (Krishnaraj, 1988: 892).

Feminism and science studies converged within this background. Recounting the debt of the Kuhnian revolution, Keller, in the introduction to her book *Reflections on Gender and Science* (1985), acknowledges that different organizations of knowledge and different interpretations of the world are consistent with (what we call) science. This was a welcome alternative to the view that science is autonomous and absolutely progressive – it approximates (ever more closely) a full and accurate description of reality "as it is" (Keller, 1985: 6). Keller questioned the purpose behind labelling one aspect of human experience male and another female. She studied how these labels affect the ways in which we structure our experiential world and assign value to its different domains. In the last analysis, she sought to find out how we acculturate and value actual men and women. She asserts that science has evolved under the formative influence of a particular ideal of masculinity.

Secondly, science is not a purely cognitive endeavour. It is a deeply personal and social activity. She calls this activity the 'science-gender system'. Feminism's confrontation with the scientific community enlarges our understanding of the history, philosophy and sociology of science. This is achieved by including those domains of human experience, which were previously relegated to women (personal, emotional, sexual etc.) (ibid.: 9).

Hilary Rose (1988: 59–72) has traced feminists' efforts to confront science studies. She primarily dealt with four issues: (a) why were there only a handful of women in science, (b) the recovery of Hypatia's sisters (forgotten women scientists) (c) the contestation of patriarchal science such as socio-biology or primate studies (which paved the way for the renaturalization of women propagating biological determinism around the 1970s), and (d) the feminist critique of epistemology.

In the introduction to her book *Love, Power and Knowledge,* Rose (1994: 1–27) credits feminists located in social studies for epistemological debates. She praises feminist biologists who debunked socio-biological accounts based on weak theory and inadequate and misinterpreted data (e.g., Ruth Hubbard's "Women Looking at Biology; Looking at Women"). Women in social studies employed the dialectical relationship between the production of things and the production of people. Without this dialectics, the knowledge produced by science appears abstract and disembodied. She also credits the women's movement around reproductive health issues for foregrounding the critical enquiry of scientific methods. Rose admits that the postmodern linguistic concern for knowledge has provided sophisticated means to analyse the text within which both science and its critiques are located. However, she criticizes the nihilistic tendencies of hyper postmodernism. These tendencies deny the possibility of any empowering knowledge about the social and natural worlds. She believes that science and technology have an imperializing agenda towards the real world.

Rose has cited Donna Haraway's book, *Primate Visions* (1990) as a turning point for methodological debates. She feels that primatology and the study of human origins are more open to deconstruction than any other science as they are observational and not experimental. The origin stories of human species are culturally and politically located. Rose concludes her chapter "Listening to Each Other: Feminist Voices in the Theory of Scientific Knowledge" by stating "that put in terms which have informed the feminist debates for me, it has been the new fusion of standpoint theory with situated knowledge which has both extended and revisioned feminism's truth claims" (Rose, 1994: 96).

Haraway's critique deploys the deconstructionist methods of postmodern feminists along with the realism of socialist feminism. Haraway states that 'social constructionists'[13] clearly indicate that official ideologies about objectivity and scientific method are bad guides to illustrate the making of scientific knowledge. She talks of 'situated knowledges' and the doctrine of embodied objectivity. She insists that all visions are 'embodied'. The sensory system produces a vision only out of the 'marked body' (Haraway, 1991: 188). Even feminist objectivity is about limited location and situated knowledge (and not about transcendence and splitting of 'subject' and 'object'). The only way to avoid gross error and false knowledge in social sciences is to accept the agency of the 'objects' studied. According to her, rational knowledge does not pretend to be disengaged or to be from everywhere and nowhere. It is not free from interpretation, from being represented, fully self-contained and fully formalisable (ibid.: 191).

Sandra Harding

Developing Gender for Approach

Harding criticizes empiricism of making scientific enterprises immune to critical and causal scrutiny. If we accept that science is a social activity (which is as culturally specific as religious, educational, economic and family activities) then we can critically interpret all scientific activities. Even those activities which make scientific activity possible would come under the purview of analysis. These activities include selecting problems, formulating and evaluating hypotheses, designing and performing experiments, interpreting results, motivating, educating and recruiting young people for the scientific workforce and its related support services (in families, psychiatrists' offices and laboratories) (Harding, 1986: 56).

Harding criticizes the inadequate efforts of feminists to theorize gender as a concept in a scientific manner. She believes that feminists have ignored the fact that gender differences are found at three levels – in individuals, in human activities and in symbolic systems. Unfortunately, they are not symmetrical and do not correspond to one another. There are variations according to the culture, space and time. Without such efforts, it will be impossible to sensitize scientific establishments to gender as an analytical category. Further, they will continue to cling to the dogmas of empiricism.

According to Harding, feminists are building a critique of existing science. The goal of feminist knowledge seeking is to arrive at theories which accurately represent women's activities as fully social. This would foreground social relations (between the genders) as a real component in human history. This project can be defended on the traditional ground of 'objectivism'. It is essential to formulate a women-centred hypothesis (as opposed to the prevailing men-centered hypothesis) in order to comprehend the gender-free one. Feminists are not trying to substitute one set of gender loyalties for another. Ultimately, they want to arrive at a hypothesis which is free of gender loyalties.

Harding points out that some proposals made by feminist theorists (generally regarded as postmodernist) are a part of the post-Kuhnian discourse and can be aknowledged as a new way of thinking. They include the concepts of knowers (the researcher), the world to be known and the process of knowing as a part of the project of inquiry. Feminists could use this method to analyse the domination of nature (by humans/men) as a factor causing the ecological crisis and supporting the modernization process. Dualistic rationality and self-formation are distinct processes in men and women. Similarly, both relate differently to nature and non-human nature. The method of knowing (developed by feminists to analyse gender in the construction and representation of science) could be extended to include human domination over nature. The sensuous, experiential ways in which women carry out their reproductive activities and build knowledge about complex, interconnected processes should be included in the construction of science. Feminists must reveal that the androcentric (man-centric) view has dominated the construction of knowledge thus far. Harding claims that feminists have felt the need to inquire at a deeper level and with different analytical frameworks (such as epistemology, psychoanalysis and philosophy) into the formation of a masculine self (ibid.).

According to Harding, these efforts can be identified as standpoint (vantage point) epistemologies. They are also called 'successor science', i.e., science as it should be developed (ibdi.: 142). These efforts are distinct from the efforts of feminist empiricists who want to modify 'bad science' by making it more refined within the modernization paradigm.

Nancy Hartsock

In the chapter "The Feminist Standpoint: Towards a Specifically Feminist Historical Materialism", Nancy Hartsock (1983: 258–272) suggests that women's lives reveal a particular and privileged vantage point on male supremacy. This can ground a powerful critique of the phallocratic institutions and ideology which constitute capitalist patriarchy. By using the structures which define women's activity (as contributors to subsistence and as mothers) and the sexual division of labour one could begin constructing a feminist standpoint. This standpoint could be used to ground a specifically feminist historical materialism.

Hartsock puts forward a schematic and simplified account of the sexual division of labour and its consequences on epistemology. These are attributed solely to institutionalized social practices. She also discusses the use of the term 'sexual division of labour' rather than 'gender division of labour' to stress the inseparability of the effects of "Nature and Nurture" and "Biology and Culture". Thus, the fact that women bear children is not a social choice. However, it is society which makes women (and not men) rear children. This choice is structured by compulsory heterosexuality and male dominance. Another reason to use the term 'sexual division of labour' is to hold on to the bodily aspect of existence. While moving towards a theory of the extraction and appropriation of women's activities and women themselves, it is assumed that there are some things common to all women's lives in Western class societies.

Hartsock has used the Marxian analogy of the worker's relationship with the capital/capitalist. Just like the worker's actions on the external world change the world and the worker's nature, "a new life changes the world and consciousness of the woman. Woman's experience in bearing and rearing children involves unity of mind and body more profound than is possible in the worker's instrumental activity."

Hartsock uses the object-relations school of psychoanalytic theory put forth by materialist psychology. According to this theory, the process of differentiation from a woman reinforces boundary confusion in women's egos and boundary strengthening in men's. Individuation is far more conflictual for males because both mother and son experience the other as definite 'Other'. The experience of oneness on the part of mother and infant seems to last longer with girls. Girls (because of female parenting) are less differentiated than boys. They are more continuous with and related to the external object world. They are also differently oriented to their inner object world. Boys (with a simpler oedipal situation and a clear and early resolution) have repressed ties to another. As a result, women define and experience themselves relationally while men do not.

The epistemology and society constructed by men are similar to the society and ideology imposed by commodity exchange. The separation and opposition of social and natural worlds, of abstract and concrete, of performance and change, the efforts to attach greater importance to the former of each pair and the reliance on a series of counterfactual assumptions are similar to the exchange abstraction. Hartsock quotes Hegel and says that masculine experience when replicated as epistemology leads to a world conceived as (and in fact) inhabited by a number of fundamentally hostile others whom one comes to know by means of opposition (even death struggle) and yet with whom one must construct a social relation in order to survive.

Women's self-construction (in relation to others) leads to an aversion to dualism of any sort, a valuation of concrete everyday life, a sense of connectedness and continuity with other people and with the natural world. Women's experience inverts that of men. It exposes abstract masculinity as partial and fundamentally perverse, as occupying only one side of the dualities it has constructed and reversing the proper valuation of human activity. The most dramatic characteristic of masculine experience is the substitution of death for life. The masculine preoccupation with death supports the argument that it is the ability to kill (and for centuries the practice) that sets humans above animals. She cites Simone de Beauvoir in claiming that "it is not giving life but in risking life that man is raised above the animals: that is why superiority has been accorded in humanity not to the sex that brings forth but to that which kills" (cited by Hartsock, 1983; de Beauvoir, 44).

The Marxian theory went beneath the surface to discover the different levels of determination which defined the relations of the capitalist with the male worker. These levels of determination and laws of motion (or tendency of phallocentric society) must be analysed on the basis of female experience. Feminists have barely started re-evaluating the female experience. They are searching for a common thread which connects the diverse experiences of women and are working for the structural determinants of these experiences.

It is imperative to generalize the potentiality of women's activities – to define society (as a whole) as a propertyless producer of use values and human beings. Generalizing the human possibilities (in the life activity of women to the social system) would create the possibility of a fully human community. This community would be structured by a variety of connections (rather than mere separation and opposition). Women's life activity forms the basis of a specifically feminist materialism which can provide a vantage point to critique and work against phallocratic ideology and institutions.

Clearly, Hartsock is a socialist feminist. However, her attempt at a 'masculine' epistemic critique (critique of the process of knowledge construction) is very important. Her standpoint theory is regarded as a milestone for Marxist feminists who believe in the material origins of women's exploitation and oppression. Marxist feminists are not as critical as Mies and her colleagues towards 'modernism' and 'rationality'.

Evelyn Fox Keller

The physicist Evelyn Fox Keller is influential among feminist scientists and social scientists. She used the object relations theory of psychoanalysis to understand the history of the philosophy of science. She traces a masculine domination of the philosophy of science back to the Greek philosopher Plato. Keller attempts to understand his method of constructing knowledge (Keller, 1985: 22–26). According to her, pre-Platonic Greek natural science was based on the principle that the world of nature is permeated by the mind. The presence of mind in nature was regarded the source of regularity or orderliness in the natural world. Its presence made a science of nature possible. Plato constructed the mind-body divide. He wanted to extricate the mind from the irrational impulses of the 'body'. Interestingly, he believed that the mind discovers knowledge only when it is guided by Eros (sexual passion). Sexual passion was interpreted as experiencing 'energy'. But, the Eros during heterosexual activity does not produce knowledge because this is an Eros which is directed towards procreation. Homosexual activity carried out in a similar manner does not take the participants to the stage of spirituality. This stage is essential for the soul to give birth to knowledge. He described a particular method of the homosexual act whereby two men face each other and avoid the position of dominant or subordinate (as observed in anal copulation). The copulation involves an older man who loves the younger man. There is dignity in this relationship. The younger man does not get aroused and does not share a 'desire' with the older man. Plato calls this 'legitimate Eros' premised on mutual consent and respect. This model of reciprocity – what Plato calls pederastic love (love between older and younger partners) – is sensual in its resonance but denies consummation. It transmutes physical excitement into imaginative and intellectual energy. Keller cites illustrations of this kind of Eros on Vases of that period. This kind of pederastic love (still hierarchical but free of domination) provides a metaphor for a form of knowledge which is erotic but continues to suffer from internal inequalities. A disregard for the embodied individual in and of itself pervades Plato's entire philosophical system, especially his theory of love. Similarly, nature remains divided between its material and logical essences with the former described as 'obeying' the latter.

Keller has described the intellectual history of two competing philosophies (hermetic and mechanical) in the 17th century. Bacon represented the mechanical view and developed the concept of the "Art of Mastery and Obedience of Nature" at the onset of industrialization. Keller asserts that Plato and Bacon did not invite material nature into a partnership of love. Plato relegated her (nature/feminine) to another realm while Bacon attempted to seduce and conquer her. Bacon inherits from the Platonic world view but incorporates the aggression which Plato had excluded. According to Bacon, the aim of science was to control and dominate nature.

Bacon believed that Aristotle, Plato, Galen and Hippocrates gave birth to an older science which "represented only a female offspring". He contrasted this to the masculine birth which he himself heralded. The aggressively male picture of Bacon's scientist could have been driven by the need to deny that the scientific mind must be (at some level) a hermaphroditic[14] mind. She calls this a child's fantasy which reveals the bisexual character of the oedipal project (an identification with the mother causing jealousy towards the father). Keller asserts that today scientists have assumed the procreative function which Bacon reserved for God. His mind is single entity – both phallus and womb. However, the scientist's kinship with Bacon survives in his simultaneous appropriation and denial of the feminine (ibid.: 38–42).

Keller has presented Farrington's analysis (1951: 197, cited by Keller, 1985: 48) of hermetic and mechanical philosophy. According to him, Bacon's root image was a "chaste and lawful marriage between Mind and Nature that 'binds' nature to man's science and makes her his 'slave'". His emphasis was on the disjunction of Mind and Matter (and ultimately on domination). In contrast, the root image of the alchemists was coition (the conjunction of Mind and Matter) – the merging of male and female, the depiction of a hermaphroditic union, sexual union or simple collaborative efforts of man and woman. Two alchemists – Paracelsus and later Thomas Vaughan – were concerned with wholeness. They believed that men and women together form a whole. Similarly, all elements should be united in a harmonious whole – an "invisible union to God". They sought the method of true knowing – the means by which one could penetrate the "innermost core of nature". They favoured experiments over theory. The masculinization of science took place along with its institutionalization in the Royal Society. The Secretary announced that the intention of the Society was to raise a masculine philosophy to ennoble the Mind of Man with the knowledge of solid truth. Even though alchemists were not feminists, they were in awe of women's procreative powers. This was a matter of envy and reverence to them. Otherwise, they shared the general patriarchal contempt for women.

The witch hunts in Europe were supported by the cosmology of Baconian science. His world view asserted that a scientist is capable of giving birth to knowledge. No sexual passion was necessary to obtain knowledge. To accomplish this purpose, women were portrayed as satanic. They were eliminated from the sphere of knowledge production. She feels that this promoted the simultaneous vanquishing of nature and female voracity. Cartesian atheism later relegated nature to the status of a machine (Keller, 1985: 59).

The alchemical concept of wholeness is similar to the Indian concept of God as Ardh Nari Nateshwar who is half male and half female. It is a concept from Tantrik philosophy. However, in Vedic philosophy, the notion of emancipation suggests that the female was a destructive enchantress who diverted man's attention from *moksha* (salvation). Overcoming sensual stimulation was vouched as the true path to liberation, i.e., obtaining knowledge.

Genesis of Methodology of Science

Carolyn Merchant (1995) has analysed the organic link between rising industrialism (supported by capitalist economy) and the ideological framework of the scientific philosophy of Francis Bacon.

Bacon is eulogized as the originator of the modern research institute, a philosopher of industrial science, the inspiration behind the Royal Society (1660) and the founder of the inductive method (using which people can verify the truths of science). He was extremely influential and aware of the important developments of his time. His language, style, nuance and metaphors mirror the class perspectives of that period.

Furthering the social transformations which had reduced women to psychic and reproductive resources, Bacon used language to reduce women to economic resources. The imagery used by Bacon to delineate his scientific objectives and methods is derived from the courtroom. It treats nature as a female to be tortured through mechanical inventions. This is strongly reminiscent of the interrogations of witch trials and the mechanical devices used to torture them. Merchant has highlighted the language used by Bacon. "The new man of science must not think that the inquisition of nature is any part interdicted or forbidden." Nature must be "bound into service" and made a "slave". She must be "moulded" by the mechanical arts. The "searchers and spies of nature" are to discover her plots and secrets (Merchant, 1995: 81).

Bacon transformed the magical tradition by asserting the need to dominate nature. This was intended not just for the benefit of the individual magician but for the entire human race. The natural magician had seen himself as operating within the organic order of nature. He was a manipulator of parts within a system and brought heavenly powers down to an earthly shrine. Bacon extended this idea to include the recovery of power over nature which was lost when Adam and Eve were expelled from paradise.

The method of 'interrogation' was not through abstract ideas but through the instrumentation of understanding. In the words of Bacon, "The instruments of the mind supply suggestions, those of the hand give motion and aid the work." "By art and the hand of man," nature can be "forced out of her natural state and squeezed and moulded". In this way, "human knowledge and human power meet as one" (ibid.: 83).

Bacon has used bold sexual imagery to depict the binding of nature in the laboratory, her dissection through hand and mind and the penetration of her hidden secrets. He ascribes technical discoveries to modern experimental methods. These new methods have the power to conquer and subdue nature – to shake her to her foundations. He writes that under the mechanical arts, "nature betrays her secrets more fully … than when in enjoyment of her natural liberty" (ibid.: 83).

Merchant concludes that "the interrogation of witches as a symbol for the interrogation of nature, the courtroom as a model for its inquisition and torture through mechanical devices as a tool for the subjugation of disorder were fundamental to the scientific method of power. For Bacon, sexual politics helped to structure the nature of the empirical method that would produce a new form of knowledge and a new ideology of objectivity seemingly devoid of cultural and political assumptions." The most useful arts to construct knowledge were chemistry, anatomy and mathematics while the best instruments included the microscope, telescope, thermometer, barometer and air pump (ibid.: 85).

Merchant notes that the concurrent artistic depictions of nature (in statues and paintings) changed from an Mother Earth whose penetration was not sanctioned to a coy female removing her own veil. From active teacher and parent, nature had become a mindless, submissive body. The new conceptual framework of the Scientific Revolution was a mechanism. Its norms were different from those of organicism. Organicism was view in the philosophy of biology which meant that each part is a whole and yet part of the greater whole, superbly coordinated (Duraiswamy, 2009: 335–363). Mechanical order and its associated values of power and control mandated the death of nature.

The Critique of the Dualist Rationalist Tradition

Descartes paved the way for the epistemology of Enlightenment. All environmental philosophers (of the 'critique of industrialism' stream) have criticized his rationality of completely severing the relationship between human and non-human nature. Though dualist rationality was present even in Plato's philosophical literature, Descartes perfected it by conceptualizing a complete divide between mind-body and reason-sensuous experience.

Plumwood has described this process: the first step involved the construction of a new normative human identity. 'Mind' or 'reason' were glorified while the entire spectrum of all other human and non-human characteristics were inferiorized. The second step was constructing the notion of the mind as exclusive of and in opposition to nature. The third step was the construction of nature as a mindless, inert entity. This dualism inspired confidence and power in controlling nature which became possible with the rise of technology and capitalism. Nature was turned into a commodity and a resource. This marks the origin of a mechanistic understanding of nature and bodily processes. Such an understanding made it possible for humanity to aspire to have power over nature. Power was achievable through the knowledge of nature. Knowledge thus becomes a source of power.

This way of thinking was internalized into distinct epistemologies when Descartes shifted the basis of mind from rationality to consciousness. According to Plumwood, Descartes made an exposition of consciousness which rests only in the mind, i.e., in thought processes. He turned logic upside down assuming that those who did not have a mind did not possess consciousness. With one sweep, he discarded the whole non-human world as non-agentic and directionless. This justified treating animals like instruments.

Plumwood suggests that this epistemology had a profound effect on women. When production was separated from the household, upper class women were perceived as civilized, domesticated and as having tamed natural instincts. In contrast, colonized women's nature and that of the lower classes was seen as savage. In the pre-industrial period, women were perceived as either Mother Goddesses or wild, chaotic temptresses. Now, they became the 'other' – passive housewives closer to nature.

The loss of continuity with nature and non-human nature is the result of a self-account which is implicitly masculine. Plumwood argues that the Cartesian knower – the 'thinking substance' – lacks a body and is unlocated. His knowledge is not limited by the bodily perspective. In the Cartesian account of achievement and identity, the body is backgrounded and denied. It is associated with the feminine and with marginalized groups ('primitives', 'animals', slaves and those who do manual labour). This is how the 'Master Self' is created (Plumwood, 1993: 104–119).

The key elements used by ecofeminists to point out the dualist approach within the existing paradigm of knowledge/science are listed by Plumwood in the following manner. She admits that this is not an exhaustive list. Other feminists continually add elements to it to make it more relevant and inclusive:

> Culture/nature, Reason/nature, male/female, mind/body (nature), master/slave, reason/matter (physicality), rationality/animality (nature), reason/emotion (nature), mind (spirit)/nature, freedom/necessity (nature), universal/particular, human/non-human, civilized/primitive (nature), production/reproduction (nature), Public/private, subject/object, self/other. (ibid.: 43)

Greta Gaard adds her own concerns to the list, e.g., white/non-white, financially empowered/impoverished, heterosexual/queer (homosexual) and reason/erotic (Gaard, 1997: 116). These elements have hierarchical positions. Whatever is equated with nature is considered inferior. Ecofeminists insist that all daily life activities are organized and legitimized based on this ideology. These ideas define the paradigm within which future knowledge gets generated in physical, natural and social sciences. Plumwood quotes the author Eve Kosofski Sedgwick, who finds these normative dualisms subsisting in an unsettled and dynamic tacit relation. The instability is caused by the fact that the second term of the dyad is constituted as at once internal and external to the first term. Her explanation is compatible with the linking postulates designed by Plumwood in the following manner:

1. Backgrounding in which the master relies on the services of the other and simultaneously denies his dependency.
2. Radical exclusion in which the master magnifies the differences between self and other and minimizes the shared qualities.
3. Incorporation in which the master's qualities are taken as standard and the other is defined in terms of her possession or lack of these qualities.
4. Instrumentalism in which the other is constructed as having no ends of her own, her sole purpose being to serve as a resource for the master.
5. Homogenization in which the dominated class of others is perceived as uniformly homogeneous (Plumwood, 1993: 42–56).

These postulates can be used to analyse many situations of subtle and overt domination under the modernization regime. If they are to be avoided in daily practice and in the activity of knowledge construction, then there is a need to develop an ethical understanding of the relationship between the sets of elements which constitute our life. Discussion on ethical discourse touches the very questions of where and how to locate oneself vis-à-vis other humans, non-humans and nature without being oppressive and dominating without being subsumed and dissolved.

Defining Feminine Self: Theory of Sexual Difference, Critique of Psychoanalysis

French feminist psychoanalysts debunk dualist rationality as the basis of scientific enquiry. E. Grosz has presented Irigaray's work in her book *Sexual Subversions* (Grosz, 1989: 100–183). Irigaray criticizes feminists who use the same dualism and claim that by considering the woman as the 'other' her capacity for reason was suppressed. She challenges the theory that binary opposition is responsible for the genderized psyche of a person. Irigaray states that women are subjugated not because their 'masculine' traits are suppressed. Rather, they do not explore their own femininity and their own sexuality. They are unable to explore because their 'self' is dependent on men and is defined by men through opposition. Her arguments resonate with the ecofeminist assertion that femininity is a distinct construct. It cannot be understood as the inferior partner in the binary opposition of masculinity/feminity. In her theory of sexual difference, she asserts that there is a strong association between feminity and maternity/motherhood. This supplies 'energy' to women and to their 'self'. Irigaray's argument supports ecofeminism because she criticizes empiricist feminists of devaluing maternity in the name of reproductive freedom.

She questions the conceptualization of the psychic structure of mankind as it relates to sexuality. Many theories of modernization have been modelled on this structure. Freud's model of human subjectivity represents all variations of subjectivity according to a singular (Western, capitalist, white, Eurocentric) male model. In her theory, she uses deconstruction to analyse representations of both sexes to understand their 'sexual self'. She is concerned about asserting an 'autonomously conceived female sexuality'.

In Irigaray's construction of sexual difference, the concept of sexuality is based on 'corporeality' and 'morphology'. It discusses the body and anatomy of each sex which is lived by the subject and is represented in culture. Morphologies are the effects of the psychological meanings of the developing child's sexual zones and pleasures. They are meanings communicated through the hierarchical structures of the nuclear family as well as the effects of a socio-symbolic inscription of the body which produces bodies as discursive effects.

In every science (natural or social), women are either suppressed or their experiences are erased. There is a need to epistemologically explore all areas of knowledge construction. Later, these critical feminist processes need to be included in the mainstream history of scientific philosophy. There is paradigm change possible in all sciences. Women's Studies itself as a social science is the product of a paradigm change started by feminist scholars. Ecofeminism points out the inadequacy of these efforts thus far. It aims at introducing the 'subsistence perspective' as a paradigm change to disengage from the colonizing effect of the world capitalist system.

8

Ethics in the Context of Ecology

> The master model of egoism and instrumentalism, in which the systems of class, gender and colonial oppression have a heavy and connected investment, has been immensely influential in shaping the way the human relation to nature has been conceived in the west. The dominant traditions of western culture have viewed relationship to plants, rivers, animals, places and ecosystems as entirely instrumental, and defined human relations to others in nature in the same terms as the egoist defines his relation to other – humans stand apart from a nature conceived only as a means to satisfy essentially self-contained human interests. (Plumwood, 1993: 147)

A major criticism of the current knowledge building process is that the identities of man, woman, human, nature and non-human nature are formed under patriarchal domination. Hence, there is an urgent need to reconstruct these identities and their linkages in a non-exploitative manner. These definitions are important while dealing with ethical questions. One such question is how to deal with non-human nature (as it does not have a voice). Ecofeminists insist that this issue should be discussed at the level on which we formulate concepts about self-humanity-nature.

Non-human nature is located (along with us) on a continuum. We need to ascertain our ethics before dealing with non-human nature. This question should be posed within an ecocentric world view because it interrogates how we ought to relate to those 'others' who are a part of nature. Ecofeminists assert that the question of relating to non-human nature should not be confused with traditional morality. It is a part of the larger philosophical question of conceptualizing our ethical premise.

Ethics for the Use of Nature for Production: Nature as Living Organism

Carolyn Merchant (1995) states that we can deal harmoniously with nature by using the principles of ecological science as the basis of production. While tracing the history of present scientific methodology, Merchant avers that holism was first proposed as an alternative to the mechanistic view of the world by Smuts in 1926. Merchant opines that this holistic approach is provided by ecological science. Holistic presuppositions about nature have been adopted by communities which live in equilibrium with their environment. The idea of cyclical processes, of the interconnectedness of all things and the assumption that nature is active and alive are fundamental to the history of human thought. No element of an interlocking cycle can be removed without disrupting the entire cycle. The parts themselves take meaning from the whole. Each part is defined by and dependent on the total context. The cycle reflects the dynamic interactive relationship of all its parts and processes as a dialectical relation. Ecology necessarily must consider complexities and totality. It cannot isolate the parts into simplified systems because such isolation distorts the whole.

Merchant believes that feminism and ecology are interrelated because both seek common policy goals. She has formulated simple principles as under which ought to be kept in mind while dealing with nature (Merchant, 1995: 46–148):

1. *All parts of the system have equal value*

Ecology assigns equal importance to all (organic and inorganic) components in the structure of an ecosystem. Healthy air, water and soil (non-biotic components) are as important as the entire range of biotic parts including plants, animals, bacteria and fungi.

Merchant notes that the feminist arguments run parallel to the above principles. Feminism asserts the equality of men and women. Intellectual differences are not gender or race specific. The lower position of women stems from culture rather than nature.

Thus ecologists and feminists assign equal value to all parts of human-nature system. In case of ethical conflicts, each case is discussed from the perspective of the interconnectedness of all parts and the good of the whole.

2. *The earth is home*

The earth is a habitat for living organisms. Ecology is the study of the earth's household. For ecologists and feminists, the earth's house and the human house are to be cherished. Hence, chemicals and forms of energy which sustain life are acceptable. However, those which endanger life have to be resisted. Both streams are concerned about products which are hazardous to the environment.

3. *Process is primary*

The first law of thermodynamics (which is also the first law of ecology) asserts the conservation of energy in an ecosystem – energy is changed and exchanged in its continual flow through the interconnected parts. The world is active and dynamic. Its natural processes are cyclical and equipped with feedback mechanisms.

Emphasizing the dynamic processes of nature has implications for change in human societies. It encourages an open discussion of all alternatives in which ecologists, technologists, lawyers, workers, women and men participate as equals.

4. *There is no free lunch*

"No free lunch" is an essential law of thermodynamics. Energy (in the form of work) is needed to produce organized matter. Each step up the ladder of organized life, each material object produced and each commodity manufactured increases the entropy of its surroundings. Hence, the reservoir of energy unavailable for work increases. Whatever is taken from nature must be returned through the recycling of goods and the sharing of services.

Feminism believes that no free labour should be available for reproduction. Thus, the dualism of separate public and private spheres should be severed thereby merging the male and female roles in the household and the workplace. Men and women should engage together in the production of use-values. They should work together to scale down the production of commodities that are costly to nature.

Historical Materialistic Conceptualization of Ecology

Robert Goodin's theoretical work (1992: 19–83) assesses the consistency and cohesiveness of the ethical arguments provided above. It is useful as a guiding political principle for the green political arena. In his book *Green Political Theory*, Goodin offers a detailed materialistic conception of ecological reality. This forms the basis of the values/ethics of a society facing ecological questions. His concern is that green politics lacks a coherent liberatory theory. Hence current politics is nothing more than an alliance of convenience between different interest groups (affected differently by the assault on nature). He searches for a coherent thesis on the premise that the key aspects of environmental critiques are centred on the way in which the control and exploitation of nature contributes to the control and exploitation of humans.

Goodin asserts that in a market economy, the losers are humans and non-humans without market power. Their voices are not heard. He envisions the convergence of environmental issues with issues of justice to the poor. Environmental degradation is assumed to have begun with settled agriculture. Since the pace of development was slow, the damage was not noticeable. The market economy and population growth accelerated the deterioration. The poor are the most severely affected as their access to basic resources (such as water and food) gets restricted. The quest for coherence is not a demand to submerge each form of oppression into a single, amorphous, oceanic movement. It merely asks that each form of oppression develop sensitivity to other forms at the practical and theoretical level. He concludes that the green theory of value advocates minimal interventions into the natural order. He uses the concept 'green theory of value' because he feels that economists will understand its importance (as they are used to the concept of value in terms of 'consumer satisfaction', 'labour inputs' and 'natural resources inputs').

Goodin theorizes the practices generally observed by a green political person. He discusses the logical integrity of these practices and brings out the uniquely coherent principles on which green politics needs to be based to encourage resource use in a certain fashion.

The 'green theory of value' states that nature is created through certain historical processes. The author argues in favour of the immeasurable inherent value of the naturalness of things. He believes that people want to see unity and coherence in the plans and projects of their lives. They wish to see these projects in some larger context outside their own life. Natural processes provide such a context. People value nature's creations because they were produced by processes larger than (or outside of) humanity. It is important that we understand ourselves and our environment as a part of nature. This kind of perception will lead to a rejection of the economy of exploitation and the uncontrolled pillage of natural resources and raw materials. In economic theory, 'land' (unlike labour and capital) is regarded as a free good. This perception can lead us away from destructive interventions on 'land' as fertile soil is a product of larger natural processes.

Goodin uses the term 'post-material values' to indicate a kind of material restraint (after acknowledgement of a basic material existence). Another principle used by environmentalists is that of 'protection and preservation of irreplaceable assets'. Goodin supports this principle after arguing against the utilitarian argument that everything is substitutable and interchangeable. He asserts that history is necessarily irreproducible and hence should be valued.

Concern for the future is a crucial component of environmental ethics. This argument specifically addresses economists who deploy cost-benefit analysis to ignore the consequences of human actions. They do not view human history in geological time and believe in discounting the future. In defence, one could claim that nature is indistinguishable and hence cannot be considered a tradable commodity. Cost-benefit analysis cannot ascribe any value by discounting the future in the process. Nothing can be considered equivalent to what is being lost.

Goodin also offers his critical comments on the personal lifestyle politics of the greens. It is acceptable for them to follow what they believe in and live their lives in harmony with nature. It is logically coherent and ethically defensible for people to have their own identities and personal projects. It is also important that people fit into larger (natural) contexts. But Goodin makes it clear that this should be a personal agenda. It should not be emphasized as a public policy/stance. It is not necessary that those who believe in public policy for the green cause should endorse all these projects and identities. Similarly, lifestyle politics need not become identical with (and therefore subsume) public policy politics. Green value theory provides a theoretical and ethical context for public policy on environment.

Construction of New Self

Ecofeminists like Ariel Salleh (1993), Roger J.H. King (1991) and Patricia Jagentowicz Mills (1991) have developed their arguments about environmental ethics in response to the exposition of male writers (including those belonging to the 'deep ecology' stream[15]). These writings deal with our relationship to 'others'. According to Salleh and Mills, male writers in pursuit of environmental sustainability focus on moral respect for nature (including non-human nature). They pursue it like a cognitive issue (a matter of moral principle conceived as ethically obligatory and pursued disinterestedly) and not through inclination.

Ecofeminists advance the argument that this view of morality (based on the concept of reason as opposed to the personal, the particular and the emotional) falls into the trap of dualism between masculinity and femininity. If we apply the same thesis to a peasant using traditional knowledge in ecosystem-based agricultural activities and complying with sustainable principles of cultivation, he will not be considered as following environmental ethics. This is so because he does not consciously and morally believe in these actions. Ecofeminists believe that there is a need to evolve a different terminology which will take into account the features ascribed to two distinct poles and perceive them as part of a continuum between male and female identity.

At present, conceptualization is problematic at three levels, viz., self, human and nature. Firstly, the self should not be defined as a masculine self. Secondly (based on a new identity of the self on the continuum of characteristics attributed to both sexes) the identity of human should be defined. This new identity should include the attributes of nature (i.e., characteristics of reproductive work, nurturance, caring, emotion, intuition, erotics etc.) which will connect humans with nature. This will encourage our attitude towards the 'other' to remain as caring as the one between mother and child. This will form the basis for a new ethics.

Carol J. Adams (1991) analyses the question of eating animals. She problematizes whether one needs to adopt vegetarianism to profess oneself as caring for nature. This discussion acquires a different intensity in America where the everyday diet consists of large proportions of meat. Hunting is another activity which is permitted in the vast wilderness albeit restricted to certain areas. This raises the issue that is it acceptable for tribals to kill animals as their ethnic right. The third point of contention is whether the use of animals in science laboratories should be tolerated or not.

Adams starts her discussion by questioning whether we define animals as having rights. Further, are these rights more sacred than the rights of humans? She points out the six aspects of this discussion (Adams, 1991: 125–145). She feels that ecofeminists are confused about their approach to these issues. According to her, there is a positive connection between ecofeminism and animal rights which is elaborated in various streams.

The first stream consists of radical feminists who adopted ecofeminism either through the study of political theory and history or from exposure to nature-based religions or through environmentalism. A majority of them support vegetarianism as an ideology because they believe that they identify with animals and hence find it cruel to kill and eat them.

Another stream of feminists points out the environmentally disastrous effects of meat eating. Meat production demands immeasurably large amounts of natural resources. Adams states that most people do not recognize this cost because in our culture production is separated from consumption and the experience of eating. The disembodied production of a tangible product is viewed as a positive indication of economy. She also cites the health hazards of eating large amounts of meat including cancer, heart disease and the lowering of menstruation age among women.

The third stream is against intensive factory farming which does not recognize animals as beings. Animals are raised in crammed conditions and injected with hormones to accelerate their growth and size. The language utilized to describe this process angers ecofeminists. The literature used in factory farming bestows impersonal names on animals (like food producing unit, protein harvester, biomachine etc.).

The fourth strand focuses on the visualization of human nature in the current paradigm. Adams feels that we have developed a strong paradigm of carnivorous animals. Man is viewed as being carnivorous which justifies killing and eating animals as natural. This is a typically patriarchal argument because social construction is completely neglected. Currently meat is presented in a dismembered form with many other foods present. This hides the reality of killing. On one hand, it is argued that man is a carnivorous animal. On the other hand, the dead bodies of animals are renamed before consumers eat them. Patriarchal ideology establishes the human/animal duality. It also posits species difference as important in considering who may be the means and who will be the end. Later it indoctrinates us into believing that we need to eat animals. She feels that here we make ontology recapitulate ideology. Patriarchy perpetuates the ontologizing of animals as consumable bodies.

The fifth aspect of this debate focuses around the hunting of animals. It is contentious because tribals (particularly native Americans whose ethnic practices are respected) hunt animals for sacrifice. Adams states that issues need to be located in their specific environments. She proposes that it is wrong to say that all killings are wrong. Killing animals in a respectful act of the appreciation of natural phenomena does not treat animals as instrumentalities. This act is called the 'relational hunt' which is a philosophy of contingency. She admits that those who resist the ontologizing of animal eating could claim, "Eating an animal after successful hunt, like cannibalism in emergency situation, is sometimes necessary, but like cannibalism, is morally repugnant." She points out that native American culture experiences their relationship with animals very differently from the way we experience it. She argues that if we want to justify hunting by citing native American hunting practices then we should also emulate their horticultural methods and predominantly vegetarian diet.

Adams resolves the dilemma by concluding that as long as animals are culturally constructed as edible, the issue of vegetarianism will be seen as a conflict about the autonomy or choice of individuals. She argues for a need to define our I-ness and whether this I-ness requires the domination of others. The ecofeminist response is not derived from a philosophy based on rights. Rather, it is based on an understanding of our relationship with non-human nature. It brings about identification and solidarity with nature.

9

Critique of Technology

Biotechnology, Genetic Engineering and Reproductive Technology

> Biotechnology is helping the process of colonization, ongoing since the industrial revolution, first through the physical occupation of land and changing the agricultural practices to suit the needs of colonizers and later by providing the green revolution technique and chemical inputs. The new regime of globalization and deregulating commerce is going on by using biotechnology to create patents. It is a process of sacrificing people's rights to create new property rights. (Shiva, 1995)

In the name of finding solutions to ecological crises (like degradation of land, scarcity of water and prevention of disease) agribusiness companies and pharmaceutical research institutions are probing the basic structures of organisms. Feminists are divided over this issue. Liberal and Marxist/socialist feminists highlight the progressive aspect of these technologies. They feel that the misuse of these technologies can be avoided by providing proper legal structures. Ecofeminists object to these technologies and have pointed out their disastrous implications. Vandana Shiva criticizes feminists who do not scrutinize these technologies for two reasons: (a) the feminist critique of dominant science and technology has not developed an adequate ecological concern, and (b) feminist theorizing (especially the stream which upholds the view that nature, diversity and gender are merely social constructions) has converged with the dominant reductionist philosophy of genetic engineering.

'Genetic engineering' and 'biotechnology' pose technical insecurities and ecological risks. The socio-political and ethical dimensions of these technologies need to be explored. Works by Ruth Hubbard (a biologist), Vandana Shiva (a physicist), Rosalyn Diprose (a philosopher) and some other ecofeminists have been used as reference. Mies' critique of reproductive technology has been used to analyse the right of women to take decisions about their regenerative powers.

Genetic Engineering

Genetics probes deeply into the identity of individual human beings. Any account of modern genetics describes how a cell's genetic content determines its metabolic processes and physical appearance. Geneticists opine that genes are ultimately responsible for a body's functioning and appearance and therefore its identity and difference.

Rosalyn Diprose is critical of the practical application of genetic theory. The potential of designer bodies or gene therapy (i.e., the 'insertion', 'modification' and 'substitution' of genetic material to correct 'defects') is a common cause for concern. Genetic theory informs a wide range of medical and social practices. Diprose adds that immunological theory informs the approach to diseases like AIDS. Genetics is also used to screen defective individuals like criminals (through DNA fingerprinting etc.) or a foetus carrying a genetically transmitted disease. Reproductive technology, biological weapons and the selective breeding of flora and fauna (for agricultural production) owe their vigour, splendour and status to modern genetics (Diprose, 1995: 165–66).

Genetics is a systematic description of the hereditary mechanism. According to Ruth Hubbard (1995: 38), genetics was invented during the period when ideologies of hereditarianism and individualism were dominant. These ideologies influenced Social Darwinists. Donna Haraway mentions that the Second World War altered our way of thinking about cellular information and relationship inside organisms (such as information, coding, control etc.). This new perspective was more mechanistic, reductionist and hierarchical than the previously favoured biochemical perspective. DNA is currently the central dogma. According to her, the term 'information' is not a good metaphor to understand the function of DNA. It does not correctly represent the contribution of DNA to protein synthesis. The entire process is more complex than it is made out to be. Genetics exaggerates the control genes exert over metabolism and development. She feels that the manipulation of genes will not achieve the biocratic dreams nurtured by scientists.

Regine Kollek states that genetic engineering technology (GET) excises individual genes or a selection of chromosomes from a particular genome. These are then transferred to a different cell and thus a different genomic background. This technique removes the barrier which normally limits the arbitrary cross-breeding of different species (Kollek, 1995: 97). In nature, cross-breeding is possible only in closely related species.

Within a relatively short period of time this method can effect several different changes. This creates a time lapse effect relative to the normal evolutionary process. For the first time in the history of life, it has become possible to design and synthesize new genetic material without needing older, related genetic predecessors. Parameters (like space, time, biology and natural history) which influenced the characteristics of individual organisms and the way they interact with other organisms are therefore altered by GET (ibid.: 98). Organisms with completely new genetic information and characteristics are being created. Novelty is the basis of investigation into potential usefulness and application.

Biotechnology

Shiva notes that biotechnology claims to increase productivity in agriculture and the animal industry. Previously, cows were injected with hormones to increase lactation. Currently, efforts are being directed towards genetic modification to achieve the same purpose. Pigs are also manipulated and bred like machines (Shiva, 1995).

A major technological development is the creation of new varieties of seeds to produce herbicide resistant crops. Efforts are under way to produce seed varieties which are resistant to diseases caused by a specific herbicide (sold by certain companies). In effect, the sale of that herbicide will increase along with the seeds of the same companies. Another effort is to introduce new varieties of seeds which have been injected with genes of organic pesticides via bacterium. These seeds are assumed to be resistant to insects, plant diseases and viruses.

Such plants may cause rapid pest adaptation. In this scenario, the pest-elimination efficacy of these plants will be lost. Farmers will be forced to revert to chemical pesticides. More chemicals will be used at a higher cost which will further damage the environment thereby posing greater risks to food security. To recover the high costs of research, MNCs will eventually promote the same kinds of seeds and pesticides which will propagate monoculture. Currently, all research is controlled by MNCs who are directing approximately one-third of their research money on pest-resistant plants.

Biotechnology is based on genetic engineering and goes beyond the green revolution (which was mainly based on hybrid seeds, chemical fertilizers, chemical pesticides and plenty of water). Environmentalists have shown that high yielding varieties (HYVs) were not really high productivity seeds. Rather, they were highly responsive to external inputs. This basically translated into increased sales of the aforementioned external input items. Genetically engineered seeds will also be more dependent on external inputs as they have been reconstructed in laboratory conditions.

Ecological Risks

According to many environmental scientists, including feminist biologists, the consequences of intentional or accidental release of genetically modified organisms (GMOs) are likely to be disastrous. They will also have a substantial socio-political impact. This socio-political impact is not on the agenda of regulatory bodies.

The proponents of these technologies argue that humans have been using some of these technologies for ages. Common illustrative examples include using bacteria to ferment barley into beer and to make curd out of milk. Even the use of enzymes to make cheese is an old practice. Traditional breeding methods too involved genetic changes.

Biotechnologists suggest that molecular-level intervention will make GMO behaviour more predictable than the products of earlier techniques (like HYV seeds). Earlier techniques improved strains of useful organisms by using directed selection from natural variation. The industry suggests that GMO offer similar or greater improvements. GMOs embody precision control while retaining proximity to natural processes.

Environmentalists doubt whether GMOs will remain under someone's control. In all probability nature will assert her authority. They also question whether GMOs will be under public or institutional control. The pace of development and the scope of modern technologies to be commercialized contribute to a qualitatively higher risk. Since the organism has not undergone the evolutionary process, one cannot predict its impact on other organisms in the environment. Its internal stability is also dubitable because of the time and space lapse. The impact of these organisms will be irrevocable. One can also question our ability to technologically manage risk.

A good example of how companies like Monsonto are pushing GM crops in India is BT cotton. Meena Menon has compiled data on the BT cotton debate and related practices in her report "Organic Cotton: Reinventing the Wheel" (n.d. 95). In 2002, draught affected BT cotton crops in Madhya Pradesh reported 100 per cent failure. Other varieties showed an average failure of just 20 per cent. She cites various sources in claiming that the story of transgenic cotton in India is marked by a lack of debate and transparency. No one in the public domain knows what is being tested or what are the results of these tests. This secrecy is maintained despite strong resistance from farmers and environmentalists such as Greenpeace (Menon, n.d.: 91) (see Annexure 1).

Christine von Weizsacker warns of an impending disaster (Weizsacker, 1995: 112–120). She predicts that the production of a small number of standardized products with high development costs and their wide distribution will eradicate diversity. Diversity is the essence of ecological sustainability and future options. Such a bio-disaster will require further manipulation which will establish a vicious circle. Alternative solutions may not get funds. She questions how one can ensure that political decisions acknowledge the voices of different future concerns when market interventions are so pronounced.

Epistemological Dimension

The aforementioned risks are based on the assumption that genetics is constructed today as a mechanistic reductionist science. Hence, there are many unresolved gaps which are overlooked in the process of creating saleable products.

Shiva defines reductionism as the belief that the world is made up of atomised fragments which associate mechanistically to make larger systems. Individual properties of the constituents are the cause and the properties of the whole are the effects. Reductionism thrives on simplicity as internal relations are ignored to determine properties and processes (Shiva, 1995: 268).

To Shiva, 'mechanistic' means a search for building blocks (or atoms or substances or matter) which make the machine. In biology, the mechanistic model identifies genes with the living organism. The understanding is that genes make proteins, proteins make us. Genes are considered self-replicating. They are responsible for the physiological and morphological properties of life forms.

Shiva challenges this tendency to totally separate reproductive cells, i.e., the germ line from the functional body. This separation supports the non-hereditability of acquired traits. It severs the feedback mechanism between environment and heredity. The role of nucleic acid is not properly understood. Hence, it is assumed that information is transferred only in one direction (from gene to protein) (ibid.: 269). Mechanistic science considers only the cell as self-replicating. In reality the whole organism replicates. The entire organism is a self-organizing system. Genes are not transgenic organisms. They do not make proteins. Rather, they are embedded in a complex system of chemical production involving other proteins.

Ruth Hubbard (1995: 38–51) explains the relationship between phenotype and genotype (i.e., between apparent traits and genes). She insists that despite the elucidation of the molecular structure of many chromosome segments, our knowledge of the structure and the biochemical make-up of DNA (or even a specific protein) does not allow us to infer which biological function is performed by which protein in the cell. The effect of that protein on different organs (or the interaction between different organisms) is also shrouded in mystery. Hubbard asserts that the entire structure is complex. There are many factors which influence the gene's performance when any manipulation is involved.

Hubbard also identifies the importance of the position effect on the structure of cells. This effect seems to be responsible for proteins with identical or similar biochemical properties having dissimilar biological functions in different organisms. The biological effect of the gene is changed when it is introduced into a different environment. She further cites the epigenic phenomenon which reveals that during cultivation (under experimental conditions), cells of differentiated tissues increasingly lose their differentiated functions. The stable inheritance of the differentiated state of a cell thus depends upon the specific organization of the tissue in the environment, i.e., in its epigenic context.

Reduction of context operates when a gene or gene product may have different biological meanings in different contexts. Hence knowledge obtained under controlled laboratory conditions becomes irrelevant when the product is put to practical use. Loss of predictability is an important consequence. The problem is that the gene is taken as context-independent carrier of information. This is a mere additive understanding.

Thus environment and time are the prerequisites and the framework for the unfolding and development of the inherent potential of a particular object. Further, they influence the object itself. They produce a specific spatial and temporal structure within this process of change.

Hubbard insists that there are limits to experimental knowledge. One can only study specific parts or aspects of nature under specific laboratory conditions. This provides a transient view of natural phenomena and natural objects. It is easy to study closed systems where limited factors are involved. Strong networks of complex systems are more difficult to discern. Unlike chemical substances, organisms created in laboratories reproduce and change further in the environment. They are experiments in the environment and with the environment. Within this context, only scientific experts have the power to define adequate strategies for resolving the resultant socio-political problems.

In biological evolution, new options arise through accidental mutations. The mutation rate is in tune with the rate of change in the organism's environment. Mutations are shuffled within the population through recombination. This continual process ultimately results in several errors and weaknesses combining with new strengths and successes.

Insistence on pursuing mechanistic count is the expression of mechanistic understanding. Shiva avers that this displays a strong historical connection to eugenics (i.e., the selective breeding of the human population to raise a 'better' lot so as to retain the 'needed qualities'). Hubbard feels that the same tendency has turned the notion of inheritance into biological inheritance. She quotes Lewontine in stating that "The molecular biology programme through simple biological systems and the analysis of proteins structure, promised a surer albeit slower way to social planning based on sounder principle of eugenics" (cited in Shiva, 1995: 270).

Socio-political

Vandana Shiva criticizes the entire process of knowledge construction which is based on reductionist mechanistic science.

She condemns its use for 'development' efforts in the Third World as causing underdevelopment. She states that underdevelopment is commonly projected as the absence of modern Western science and technology systems. However, poverty and underdevelopment are created by the externalized and invisible costs of resource-intensive and resource-destructive technological processes. The livelihoods of millions depend on these resources (ibid.: 197).

Shiva's paper in the book *Biopolitics* (ibid.: 197–205) presents the following arguments against biotechnology with reference to seeds:

> A seed is the most regenerative resource. It reproduces and produces itself. Farmers have managed these genetic resources since the days of hunting-gathering. The seed which reproduces itself stays free. It is a common resource and under the farmers' control. The corporate seed (i.e., genetically engineered seed) has a cost and is controlled by either the corporate sector or agricultural research institutions. The transformation of a common resource into a commodity and of a self-generative resource into mere 'input' changes the nature of the seed and of agriculture itself. A peasant cannot afford it and becomes pauperised by being alienated from his land.
>
> The de-coupling of seed from grain changes the status of the seed itself. From being finished products, nature's seeds and people's seeds become raw materials for the production of corporate seeds. The cycle of biodiversity regeneration is replaced by a linear flow of germ plasm from farms and forests into labs and research stations and the flow of modified uniform products as priced commodities from corporations to farmers.

The successful development and sale of crop plants resistant to branded herbicides will cause a further economic concentration of the agricultural industrial market. This will increase the market power of transnational companies. Shiva argues that the relocation of seed production (from the farm to the corporate laboratory) relocates power and value between the North and South and between corporations and farmers.

It is falsely assumed that these seeds increase productivity. Currently productivity is measured in terms of output per unit input (specifically of a monoculture crop). In a subsistence farm, there are multiple crops which satisfy different nutritional needs of the farmer's family. Individual crops also have multiple uses. Rice is food only partially. Its straw is used to feed animals, cook food and to fertilize the field for next year. Nothing is wasted and the problem of disposal does not arise. Shiva argues that the productivity of traditional farming methods is not as small as it is made out to be provided that the farmer is not deprived of minimum basic resources such as water.

Usurping land (during the colonial period), the green revolution (in independent India) and the current creation of patents are several ways to sacrifice people's rights to the process of creating new property rights.

Cary Fowler insists that patents are a means of allocating ownership, assigning control, regulating access and apportioning benefits (Fowler, 1995). Till recently, there was a debate about whether the patent system should be applicable to life forms. However, the US has already approved the patent system for biotechnological products. Such decisions sacrifice people's rights as producers and consumers. The right of knowing and being innovative is imprisoned within the restrictions of patents. It is also a type of cultural reductionism that non-western forms of knowledge (in this case, agricultural practices and medical systems) are devalued. This is part of a larger genetic reductionism wherein the behaviour of all biological organisms is expressed by patenting life forms in terms of genes.

Colonialism and socio-economic-political control are evident in several aspects of this process. All MNCs producing agricultural inputs are developed in the North. They are directed by the desire to have a worldwide market. There is no concern about these products being inappropriate for small farmers (who work in highly variable and vulnerable ecosystems). The MNCs are not even aware that they might be undermining security while securing food. The indiscriminate engineering of species and ecosystems results in ecological vulnerability as their capacity to adapt is lost.

MNCs have the double standards of these companies while negotiating GATT and Intellectual Property Rights (IPRs). MNCs willingly cross species boundaries while simultaneously separating the gene from the whole. Fowler cites several examples to illustrate those boundaries disappear for MNCs when they claim the right to manipulate and own life. However, the boundaries reappear when there is a discussion about risks and responsibility sharing.

The era of biotechnological development is marked by the instrumental transformation of nature. There is an asymmetric distribution of risks and benefits and an increase in inequitable ecological uncertainty. However, no accompanying cognitive framework for risk management has emerged.

There is a mixed response from the farmers in India. Farmers from Southern India especially have organized against the merchandising of biotechnology (Annexure 1).

Argument for Bioethics

In the book *Biopolitics*, various authors have attempted to illustrate that genetic practice should be considered ethically significant, particularly when a high degree of modification and manipulation of bodily specificity is involved. Biomedical manipulation can irreversibly change a habitat. All genetic practices, such as genetic screening and genetic engineering, are of ethical import because ultimately the supporters of this science hope to recreate man himself. Opponents fear that difference will be eradicated by reproducing the self (or making the 'other' same). Environmentally-oriented scientists are concerned about preserving diversity. A number of programmes are currently mapping the human genome. Excessive confidence is being placed in a direct causal link between genes and individual behavioural differences. This could result in misguided attempts to normalize and/or isolate individuals considered inferior.

Rosalyn Diprose reveals that the word 'ethics' is derived from the Greek word 'ethos' meaning dwelling or habitat, i.e., the place to which one returns (Diprose, 1995: 162). The habitat encompasses one's habits and 'character' (their specificity or what is one's own). It creates a sense of belonging to an ethos and encourages us to take up a position in relation to others. Thus, the concepts of embodiment and ethics intersect by virtue of our spatio-temporal-being in the world. This means that the integrity of the body and its specificity should be respected.

Diprose criticizes the prevalent understanding of ethics by echoing ecofeminist concerns. She states that we understand ethics in terms of a set of principles which ought to govern behaviour. These principles are formulated and grasped by the rational mind. As these principles claim universality, the ethics evoked by them assumes that behaviour originates in a potentially unified mind housed in a broad, homogenous habitat.

Science has distanced our 'being' and the 'world' which the body inhabits. Scientific descriptions of the body are not supposed to come under the purview of ethics, i.e., there is no place for ethics in knowledge. However, recent public scrutiny of biomedical science suggests that there is a link between science, embodiment and ethics. She complains that there is rarely any explicit reference to the significance of embodiment to biomedical ethics. She claims that biomedical science and medical practice are ethical by nature because they deal with our most intimate possession – the body. Genetics has (somewhat inadequately) brought these issues into public debate. One of the important threats is the desire to double the self, reproduce it or to make the other 'same'. This would eradicate difference and the specificity of embodiment. Diversity could be reduced to ultimately establish eugenic control.

Another aspect of traditional ethics was that inherent value can only be ascribed to 'moral agents', i.e., subjects which have (in some sense) free will and reason or at least practical reason. Jesper Hoffmeyer (1995: 151) states that there have been several attempts during the last few decades to at least bring mammals under the protection of moral duties and to ascribe inherent value to this broader category. This is to be achieved in such a way that other moral agents are bound by direct moral duties towards all such moral subjects.

One has to assess the work of regulatory bodies against this background. These bodies are being established since the advent of genetically manipulated/modified organism (GMO). They evaluate various dimensions of these phenomena and commission studies to assess their impact. Regulatory bodies are supposed to be premised on 'bioethics'. Hoffmeyer states that these bodies divide concerns into three groups:

- technical risk management
- free consumer choice
- observation of bioethical principles

Regulatory bodies are criticized of deriving their value framework from industry. Thus, they inherently support the instrumental approach towards nature. Industries extend the model of industrial efficiency to animals. Ethics become the prerogative of professional experts thereby narrowing the field of scrutiny. Hoffmeyer rues that these regulatory bodies are unwilling to evaluate those systemic disturbances which are not immediately correlated. Another complaint is that the agenda for research is not based on the broad concerns of human society. It is narrowly defined by the industry which is centred on profit generation. Further, there are socio-political consequences of such research.

Some evaluative studies show that consumer choice is restricted by the portrayal of products by the industry. Environmentalists criticize these companies of employing convenient metaphors. Hoffmeyer has noted several examples of these statements by Les Devidow. On one hand, they call their products an improved form of nature, which will protect agriculture from untamed nature. On the other hand, their slogan is "fighting for a better world, naturally". Here nature is portrayed as being under benign control. The biotechnology industry hails GMOs as reprogrammed nature. They argue that their environment-friendly products will overcome the limits of chemical intensive agriculture. Vandana Shiva (1995) calls this a political battle for control over nature. Ecofeminism cautions us against the progressive myth of these technologies.

Mies' Critique of Reproductive Technologies

Modern biological research (particularly genetic engineering and reproductive technology) has given rise to new questions on the ethical foundations of these technologies. Women's rights and dignity are central to the critique of reproductive technologies. These ethical questions cannot be separated from their historical background. They are rooted in the conception of human as master and lord over nature. This lordship is justified by his rationality. Biotechnology regards humans (mainly women) as a source of organic matter, DNA and raw material which can be worked upon. They are the targets of man's control over nature.

The application of the principle of selection and elimination was perfected through prenatal diagnosis (quality control) and in vitro fertilization (IVF). Today, one can select ova and sperm according to certain quality standards. One can even isolate genes, cut up DNA, examine which chromosomes are defective, recombine and manipulate pieces of DNA and thus directly influence genetic substance. Geneticists are mapping the genetic pool of humans, animals and plants in order to discover unknown genetic 'defects'.

Maria Mies (1993) states that the long-term social, ethical and biological changes about to be induced globally cannot be fathomed without understanding the historical background of this process. The Enlightenment period brought social reforms and concepts of equality and human rights. Instead of pursuing these concepts further on social and political platforms, society is turning to technologies which fragment and invade bodies under the guise of 'individual choice'. Such technologies allow the market and profit motives to encroach and colonize newer areas of nature (especially women's bodies). She pleads that social relations between classes and genders are based on exploitation and subordination. Technological innovations within such unequal relationships intensify inequality and further exploit the groups concerned. The technical aspects of these technologies have become general knowledge due to their use for population control. This is masked as an effort to tackle the consequences (e.g., infertility) of the careless use of the contraceptives.

These technologies were strongly opposed by developed countries (especially Europe which is intimately aware of the eugenic practices of the Nazi period). Feminist International Network of Resistance to Reproductive and Genetic Engineering (FINNRAGE) was actively involved in this debate in the 1980s and 90s. The Indian critique (from the reproductive health perspective) was voiced by many feminist organizations such as Forum for Women's Health (Mumbai) and Saheli (New Delhi). Saheli fought in the court against the hazardous effects of hormonal contraceptives.

Brief History

Reproductive technology and genetic engineering are based on the principle of selecting desirable elements and eliminating undesirable ones. The distinction between white people as 'human' and blacks and browns as nearer to 'nature' along with the parallel distinction between men and women found its clearest expression in the age of Enlightenment (in the 18th and 19th centuries).

In the first half of the 19th century, sexist and racist ideology was encapsulated in the idealistic philosophical discourse on the dualism between 'nature' and 'culture'. They were given a materialistic 'scientific' base in the second half of the same century. Darwin's theory of evolution (particularly in the form of Social Darwinism developed by Spencer) posited that the 'survival of the fittest' was the selective mechanism by which 'superior' societies evolved from 'lower' ones.

Such ideas gave rise to the eugenic movement started by Francis Galton. He coined the term 'eugenics' in 1883. Galton combined the ideas of Darwin and Malthus and advocated 'selective breeding' in order to prevent the deterioration of the race. The 'fit' were to be encouraged to breed more and the 'unfit' to breed less. Galton was interested in the genetic quality of people. He promoted statistics in social research and introduced the grading system to measure people's genetic quality. By applying statistical methods to eugenics, he gave 'scientific' legitimacy to his theories as mathematical procedures and statistics were considered proof of scientific objectivity.

The eugenic movement influenced social science, psychology, behaviourism and politics. After the publication of the biological laws of heredity (discovered by Gregor Mendel in 1865) in 1901, the movement gained momentum in Britain and USA. Eugenicists applauded Hitler when he passed a compulsory sterilization law in 1933 (known as the Law on the Prevention of Hereditary Diseases in Future Generations). The British Eugenics Review hailed Hitler's Germany as a vast laboratory which was the scene of a 'gigantic eugenic experiment'. The atrocities perpetrated by Nazis on people who were considered 'unfit' (particularly Jews but also gypsies and the handicapped) brought the eugenics movement into disrepute. However, many lieutenants of this movement entered the fields of population control and family planning after World War II. There is a historical continuity between the eugenics movement (via Nazi Germany) and new reproductive technologies (prenatal diagnosis, genetic engineering, in vitro fertilization etc.) The proponents and practitioners of these technologies do not acknowledge this historical heritage.

Ethical Issues for Ecofeminists

Maria Mies' (1993) criticisms are listed below:

1. The female body's generative capacity is treated as a new 'area of investment' and profit-making for scientists, medical engineers and entrepreneurs (in a situation where other areas of investment are no longer promising). Reproductive technologies have not been developed because women need them. They exist because capital and science need women to propagate their model of growth and progress.
2. These technologies are legitimized on humanitarian grounds by those who try to sell them (e.g., to help infertile couples have a child of their own, to help women avoid bearing handicapped children, to minimize the hazards of pregnancy

and child bearing etc.). The methodological principle is to highlight the plight and unhappiness of a single individual and appeal to the solidarity of all to help that individual. The aim is to control women's reproductive capacity.

3. The argument (that only their application determines whether these technologies are good or bad) is based on the proposition that science and technology are value-free and have no bearing on social relations. Both exploit and subordinate nature, women and other people.

Sexist biases permeate new reproductive technologies and genetic engineering at all levels. Motherhood is transformed from a creative process to an industrial production process. This disrupts the relationship of the mother with the child. The entire process is rationalized, objectified, planned and controlled by medical experts. This destroys the integrity of women as human beings and as individuals.

These developments place women's reproductive capacity under a rigid and constant quality control. There is enormous social pressure on pregnant women to produce perfect children. In countries like India and China, amniocentesis (used as a sex determination test) has caused large-scale abortions of female foetuses. In industrialized countries, women are subjected to a series of pregnancy tests. If they are over 30 or 35 they are labelled 'risk-pregnancies' and pressurized to undergo amniocentesis to avoid bearing a handicapped child. Most young women are afraid of having babies without constant medical monitoring.

The anti-women tendencies of new reproductive technology are displayed in their aggressive and invasive nature. The IVF programme starts with long-term fertility monitoring and hormonal treatment (the long-term effects of which are not known). It involves laparoscopic procedures, hormonal treatment, frequent ultrasound monitoring and amniocentesis. These physical invasions generate anxieties and traumas. Mies (1993) cites Gena Corea who describes the hope and despair of women at each stage of the IVF treatment. The woman in these programmes is made a totally passive object. Therefore, these procedures are humiliating and degrading.

The invention of various contraceptives (particularly the pill) was hailed as the decisive technological innovation which would liberate women from their unruly fertility. Many cases of sterility today are the result of invasive methods of contraception (like the Dalkon Shield and other IUDs) and callous treatment by doctors.

Population control technologies inherently disregard human rights and dignity. From 1975 onwards, there has been an increasing tendency to accept coercive methods and 'final solutions' such as sterilization. India carried out a massive sterilization campaign during the Emergency from 1975 to 1977. Millions of people were forcibly sterilized. Even the Chinese population policy has led to female infanticide and foeticide.

Contraceptives which are not licensed for use in family planning programmes in the West are tested on Third World women (e.g., NET-OEN tried in India in 1984). These tests mystify the health hazards for women while masking their inherent racism. The health hazards include breast cancer, two types of uterine cancer, serious menstrual disturbances etc.

Prenatal diagnostic technology reflects sexist and racist tendencies. Amniocentesis was initially developed to discover abnormalities in the foetus. It is widely used in India as a sex determination test. Today sonography has replaced amniocentesis. This modern technology is being used to strengthen patriarchal attitudes and institutions. This test is increasingly being adopted at the rural level as it is relatively cheap and because women who already have daughters are afraid of facing their families with another one.

Annexure 1

Resistance to Merchandising Biotechnology (Guha and Martinez-Alier, 1998: 109–120)

Introduction

A few years ago, the Uruguay Round of GATT negotiations tried to force India to accept the intellectual property rights of commercial agricultural seeds. The North American Free Trade Agreement (NAFTA) may be the final blow to traditional agriculture in Southern Mexico. These two events raise general questions about the conservation of wild and agricultural diversity.

Genetic Erosion

Indigenous groups are a veritable storehouse of knowledge about biological diversity. Peasant farmers select and improve these seeds. According to Mathreyi Krishnaraj, an NGO in Bangladesh had collected over 500 types of paddy seeds under the International Fund for Agriculture Development (IFAD) programme. But this knowledge is rarely valued in economic terms. 'Genetic erosion' occurs because no one has appropriated genetic resources and treated them as merchandise (because things without a price or an owner are treated as worthless). Although round figures are often used, it is difficult to construct indicators of genetic erosion. The names of varieties, the extent of the farmer's reutilization of seeds and the genetic distance between varieties of traditional agriculture have not been recorded. Despite the lack of accurate information, there is an increasing concern about genetic erosion.

Peasant Struggles to Control Seeds

An indigenous group which is outside the generalized market system and which has ancient and proven methods and a vast botanical and zoological medicinal knowledge is not built in a single generation. This knowledge could be transferred to an outside group without anything in return. This could occur by means of scientific research, missionaries of other religions, political or economic exploitation etc. This outside group can translate and absorb this knowledge into its own culture. It can manipulate materials to apply this knowledge in its own medicinal system. In India, the Karnataka Rajya Raitha Sangha (a peasant organization), in association with the Third World Network, organized a variety of mass actions in 1993–94 against transnational seed companies to oppose intellectual property rights over 'improved' seeds. If these rights would have been implemented, the farmers would have been unable to produce these seeds and trade them among themselves.

Agricultural Biodiversity as 'Cultivated Natural Capital'?

The main aim of organizations like the Worldwide Fund for Nature is to conserve natural biological diversity. This agenda has received more attention in the strategy of the International Union for the Conservation of Nature than agricultural diversity. There is a complementary relationship between wild and agricultural biological diversity. Agricultural genetic resources can be called 'cultivated natural capital'. They cannot be fully substituted by capital goods (including 'improved' seeds) used in modern agriculture. This cultivated natural capital should be complemented with natural capital, i.e., the wild relatives of cultivated plants.

'Farmers' Rights'

Activists in favour of ecological agriculture believe that CGIAR's agricultural research centres should not patent their genetic resources. They are generally against intellectual property rights because they do not believe that these defend and reward agricultural diversity. They do not even agree with the payment of 'Farmers' Rights'. This payment cannot buy the right to exclusive use of genetic resources. Farmers' Rights should be prized and honoured. They should give an incentive to ensure the conservation and development of agricultural diversity. Rather than paying royalties for traditional seeds, all genetic resources should be considered 'world heritage'.

The INBio-Merck Deal

The deal between INBio (Instituto Nacional de Biodiversidad) of Costa Rica and the pharmaceutical company Merck reached in 1991 is a glaring example of the exploitation of the poor. INBio has free access to the conservation areas of Costa Rica. It does not pay the direct costs of establishing and guarding the national parks or the cost of maintaining these wildlife reserves. Apart from a large sum of money, Merck pays royalties on profits from any commercial products. But this limited monetary incentive is insufficient to prevent deforestation and genetic erosion.

The Defence of the Agroecology Outside the Market

A problem arises when ecological agriculture cannot compete in the wider market with modern agriculture. Previously, there was no generalized social awareness about the peasant's ecological virtues. There has arisen an international ecological and agrarianist movement which stresses the peasantry's role in the development of genetic resources. Ecological agriculture spread because of the peasant's own learning process. The spread of modern agriculture also happens this way. It is the poor farmers who must (in every season and on every site) decide whether to conserve traditional varieties or to adopt commercially improved ones. The farmers are influenced by a variety of considerations pertaining to livelihood and use value along with commercial considerations.

Annexure 2

FINRRAGE
(Corea, Hanmer, Klein, Raymond and Rowland, 1987)

Since the mid-1970s, women have organized a series of national and international meetings to discuss technologies and plan resistances. The first conference was held in USA in Amherst, Massachusetts in 1979. The Leeds group of Women's Reproductive Rights Campaign sponsored the next national conference in Great Britain in 1984. A session at the International Interdisciplinary Congress on Women (held in Gröningen in the Netherlands in April, 1984) led to the establishment of a women's network on reproductive technology. The network monitors technologies worldwide. The network members organised a conference devoted solely to reproductive and genetic engineering in Vällinge, Sweden in July 1985.

The Emergency Conference passed a resolution and changed the name of the network from FINNRET (Feminist International Network on the New Reproductive Technologies) to FINRRAGE (Feminist International Network of Resistance to Reproductive and Genetic Engineering). The basic information was first shared with women from 20 countries from the First and Third Worlds. This is a young network in which contacts and coordinators assemble from 20 countries. They sort and distribute medical information in their respective countries. The members highlight how these technologies affect all women. They arrive at a position on these issues and make critical judgements about their consequences on women. They also discuss some arguments which confuse other women thereby distorting feminist resistance to technologies.

Members of FINRRAGE observed that new reproductive technologies such as in vitro fertilization are presented to the public as therapy ("new hope for the infertile") and as benevolent means of expanding people's options. However, they are a powerful means of social control. These technologies are not confined to the infertile. For example, electronic foetal monitoring was introduced in obstetrics for "high risk" obstetrical complications. But now it is used on most pregnant women. A similar pattern emerges with respect to ultrasound, amniocentesis, caesarean section, genetic testing and counselling. The expansion of new reproductive technologies to an ever greater proportion of women leads to the reduction of babies to products produced by "technodocs" in their new industrial process. It also reduces the overall number of women ("preference" for male children leads to the elimination of female children even before they are born) and interferes with human evolution.

Hillary Rose (1994: 185) highlights the negative aspect of FINRRAGE campaigns which portray infertile women and surrogate mothers as 'victims' of "techno-patriarchy". It uses horror stories to denounce surrogacy and motherhood as forms of 'prostitution'. On the positive side, FINRRAGE tactics have put reproductive technology and genetic engineering into the political agenda. However this has been achieved at the cost of imposing a universal politics on an immense diversity of location. She also quotes Janice Raymond while drawing on Mary O'Brien's analysis as a story of men's relentless desire to overcome his alienation from his seed in the act of procreation appropriating reproduction from women. The FINRRAGE agenda has been deepened since the 1989 meeting in Bangladesh. At this meeting, the environmental implications of genetic engineering on reducing biodiversity were included. The "Declaration of Commilla" linked the concerns of North and South feminists. Its first demand was to halt genetic and reproductive engineering. Rose cites Maria Mies in claiming that the central struggle against reproductive technology is different for North and South feminism. Anti-natalist technology (coercive contraception) is mainly targeted at poor women of the South whereas pro-natalist technology is mainly targeted at white middle class women.

Rose comments that feminists associated with FINNRAGE frequently reduce women (and men) to biology in which (for this particular struggle) reproduction is the central function. She simultaneously accepts that FINNRAGE's strong attack has publicised the dangers of eugenicism.

In India doctors willingly entice women to undergo amniocentesis to determine the sex of their foetus (and then abort it if it is female). The imbalance in the sex ratio (particularly in the age group of 1–6) has been dramatic during the last twenty years. The campaign against sex selection through amniocentesis was inspired by the aggressive discussions of FINNRAGE (during the 80s in India) and led to the ban of sex selective abortions at the state and central level. Implementation of these laws is tardy and fast commercialization of medical practice is gaining ground.

10

Alternative Development

Visions and Strategies

> One of the sustainable levels of external inputs is that of water. This paradigm treats biomass, which is effectively a renewable energy in usable form, as the main provider – the central resource. If water is made available much more widely but in smaller quantities, it is capable of ensuring a livelihood for almost all rural families in India. If a crop pattern including trees is used in such a way, then the same quantity of available water can provide much more aggregate productivity than the water used for high input, high value crops. Equity, sustainability and productivity can be clearly seen to be going hand in hand. (Datye, 1997)

This chapter presents alternative visions of development which are compatible with an ethical framework. First of all, Carolyn Merchant's (1995) vision of a partnership society is discussed. K.R. Datye's (1997) strategy (based on a technological understanding of the elimination of unequal energy exchanges and the promotion of biomass based development processes) to ensure distributive justice in the Indian context is also mentioned. This strategy corresponds with Merchant's vision. Vandana Shiva (1991) has formulated the principles of a sustainable society. Maria Mies (1977, 2001) and her colleagues have contributed to this process of vision building and strategizing. Their ideas on the subsistence perspective have been incorporated in this discourse. Rajni Bakshi (2007) has assimilated Paul Ekin's ideas on ecological economics. They are relevant to understanding how this paradigm shift can be achieved.

Partnership Ethics

After describing the three forms of environmental ethics that currently dominate human-environmental relations, Carolyn Merchant (1995) proceeds to develop her own vision which she calls 'partnership ethics'. The three forms she has named are egocentric, homocentric and ecocentric. These forms of environmental ethics are exemplified by the three major interests represented at the Earth Summit – the General Agreement on Tariffs and Trade (GATT), the United Nations Commission on Environment and Development (UNCED) and the environmental organizations attending the global forum.

The three tendencies incorporate the following assumptions (Merchant, 1995: 209–224):

Egocentric

It is based on the idea of trickle down economic benefits, i.e., what is good for the individual (or corporations acting as individuals) and what is good for society as a whole. Nature comprises resources which can be turned into tradable commodities. It is an inexhaustible tap of free goods whose wastes are eliminated in an inexhaustible sink. Following the factory model, nature is conceptualized as a dead machine whose parts are manipulated for assembly line production. Resource depletion and environmental pollution are not part of profit/loss accounts. Hence, there is no accountability to or for nature. GATT's egocentric ethic promotes transnational corporations (TNCs) and limits democracy in these industries. The women's NGO treaty (adopted by the Global Forum) indicts GATT as a major cause of environmental destruction.

Homocentric

UNCED's sustainable development programme represents the homocentric ethic. It is a utilitarian ethic based on the precept of the "greatest good of the greatest number of people for the longest time". The utilitarian calculus of benefits and costs guides ethical choices. Homocentric ethics conflict with egocentric ethics through the influence of private interests and lobbyists pressing government officials.

The ethics of UNCED view nature as a resource for humans. It is a source of commodities. However, this principle brings future generations into the accountability calculus. The Earth Summit's goal is to promote greater democracy (for more people for a longer time) by developing and conserving resources in a sustainable manner.

Ecocentric

This view was developed by the ecologist Aldo Leopold who formulated the land ethic in 1940s. It has been elaborated as ecocentric (and biocentric) ethics by environmental philosophers over the past three decades. Ecocentrism includes the entire biotic and abiotic sphere. Leopold's land ethic expanded the human community to include "soils, waters, plants, animals or collectively the land". Ecocentrism elaborated this logic by asserting that all things have intrinsic worth and not just instrumental and utilitarian value. Ecocentrism expands the concept of 'good for the human community' to include the 'good of the biotic community'. From an ecocentric vantage point, accountability must include the rights of all organisms to exist. It argues that biodiversity is necessary for maintaining the present and future health of the entire biosphere, for enhancing the quality of life and for aesthetic enjoyment. Merchant reveals that the ecocentric ethic conflicted with the egocentric ethic of TNCs. These companies felled timber in rainforests and converted the leftover deforested areas into cattle grazing grounds.

The ecocentric ethic demands integrity of the ecosystem and insists on preservation even at the cost of indigenous people. The egocentric view victimizes nature in the interests of people who want to make profit. Conversely, the ecocentric view victimizes people in the interests of nature. Merchant analyses the shortcomings of the three approaches. She criticizes egocentric ethics of privileging few at the expense of many (cut-throat individualism). Homocentric ethics privilege majorities at the expense of minorities while ecocentric ethics privilege the whole at the expense of the individual (holistic fascism).

New Vision

Merchant explains that the term 'partnership' avoids gendering nature as a mother or a goddess. It does not endow males or females with a special relationship to nature or to each other (essentialism) and admits anthropogenic or human-generated environmental ethics and metaphors. A partnership ethics of earthcare allows both women and men to enter into mutual relationships with each other and the planet independently of gender. Women are not held responsible for "cleaning up the mess" created by male dominated science, technology and capitalism. Partnership ethics has four precepts:

- Equity between human and non-human communities.
- Moral consideration for human and non-human nature.
- Respect for cultural diversity and biodiversity.
- Inclusion of women, minorities, and non-human nature in the code of ethical accountability.

Vandana Shiva's principles (1991) are listed below:

1. A knowledge system with an ecological understanding of nature.
2. A technological system for processing resources to satisfy human needs with minimal resource waste.
3. A rational criteria for demarcating vital and non-vital needs and for distinguishing between resource destructive and resource enhancing technologies. Greater resource use efficiency is also demanded.
4. Increasing Gross National Product 2 (GNP2), i.e., the productivity of nature, e.g., development of better hydrological practices which increase infiltration and percolation to underground water resources.

5. Increasing GNP1, i.e., productivity of man made processes without reducing GNP2, i.e. productivity of nature (e.g., improved knowledge of organic farming).
6. People's involvement in the evolution of ecological sciences because it is imperative that they determine their own development path. Rural people are likely to have retained ecological perceptions of nature even when privileged groups have lost them.

Alternative to the Present Food Security Crisis

Maria Mies and her colleagues have developed the subsistence approach as a politics of ecofeminism. They treat food security as the major concern of women in the North and South. Women shoulder the responsibility of bringing food to the family and contribute towards reproduction of the family members, the community and the society. The subsistence approach criticizes contemporary policies – nationally and internationally. It provides an alternative framework for maintaining food security. It was articulated in 1996 at the World Food Summit at Rome organized by FAO. Many women's organizations conducted parallel meetings opposite to FAO headquarters. Maria Mies elucidated the following theses in a speech which were later published in booklet form (Mies, 1996: 30–37):

The mainstream wisdom of food security tends to represent consumers in the North (where land is not cultivated through family labour) and is characterized by the following biases:

- Increasing population pressures are the cause of poverty, hunger and environmental destruction.
- There is not enough land to produce food.
- Hunger can only be removed by economic growth and the industrialization of agriculture.
- Only international trade of agri-products can establish food security while domestic production, self-sufficiency and local markets cannot.

- Urbanization and flight from the countryside are natural laws and not the result of the destruction of the economic and ecological bases of the land.
- Biological and genetic engineering are the key to more food production and therefore to greater security.
- The global problems of women, especially poor women, are to be solved through increased participation in this economic system which is structurally hostile to nature and mankind.

Mies emphatically states her overall thesis that women are the chief nurturers of children. In developing countries, the harvest reaches the table only after it is processed, conserved and prepared into foodstuff by women. This work is invisible due to its 'informal nature'. Food is not just mere consumption of calories. It is an entire culture in and of itself. Women are the main keepers and reproducers of this culture. Food security cannot be guaranteed by destroying this culture in the face of modern homogenized industrial food products. Mies analyses that the invisiblization of women's contribution is a structural necessity of an economy based on profit maximization. She is against industrialized agriculture and food processing. Her solution involves decentralized community owning agricultural practices. Her main thesis is that "food security can only be guaranteed when people, within the framework of local and regional economies, take responsibility for the maintenance and regeneration of land, water and other vital resources, not only as consumers but also as producers; take responsibility for quantity, quality, distribution and consumption of food. Where self-sufficiency and self-provisioning and not profit maximizing and economic growth, are the main aims of economic practices there people will maintain biological and cultural diversity and therefore food security, for all."

Mies further explains that the food security of the South should not rely on multinationals and their trade practices. The dumping of the North's overproduction is harming local farming systems.

- There can be no food security without self-sufficiency and control of local resources.

- Food security cannot be realized in a global, liberalized and deregulated agrarian market. It is achievable only through decentralized food systems under the control of local communities.
- Commons should not be privatized.
- There can be no long-term food security without ecologically-sound farming methods.
- Retention and revitalization of local market structures.
- Such autonomous and subsistence-oriented economic areas (in relation to basic food supply) demand a division of labour between men and women different from the current gender division.
- It is therefore absolutely necessary that men fight actively against an increasingly brutal image of man (i.e., against the 'ramboization' image).
- The tendency towards urbanization and flight from the countryside must be stopped and reversed.
- No food from genetic engineering labs.
- No patents on life.
- We have right to know what we are eating.
- No to novel foods.
- We should resist the production and consumption of methods and technologies which destroy ourselves and nature.

Maria Mies' Principles

Maria Mies (1997) has drawn some basic principles of an alternative society:

1. Human activities should begin with the recognition that our planet is not infinite and we (as humans) are limited.

2. We need to incorporate survival ethics (which prevailed in the pre-industrial mode of production). The current tendency to externalize morality from economic sphere should be dropped.

3. The definition of what constitutes a good life has to be changed.
4. The concept of work should combine burden and enjoyment activities. The current division between alienated work time, pleasure and alienated leisure should be abolished.
5. The concept of freedom from dependency needs to be introduced in developed and underdeveloped countries for basic subsistence.
6. A large self-provisioning economy will drastically reduce non-productive work, i.e., work in the service sector. If the use of technology is reduced then a larger workforce will be required for food production. This will bring people closer to nature.
7. This will reduce the distance between producers and consumers which will help people to regain autonomy over their lives.
8. Human relationships with each other will remain non-exploitative.

Mies believes that mutually enriching gender relations will remain at the centre of social ecology in such a society. She insists that feminists should focus on social ecology which is grounded on the relations between men and women. The most fundamental demand of the feminist movement was autonomy over our lives and bodies. This is valid to the struggle for an ecologically sustainable society.

Subsistence Perspective Strategy Continues

The book by Veronika Bennholdt-Thomsen, Nicholas Faraclas and Claudia Von Werlhof, entitled *There is an Alternative: Subsistence and Worldwide Resistance to Corporate Globalization* (2001), has an entire section dedicated to papers documenting successful subsistence activities in countries of the North and South. The title "There is an Alternative" was specifically coined to oppose the

TINA syndrome (i.e., "There is no Alternative") proposed by the World Bank and mainstream economies. The initiatives described in the book deal with the active involvement of women in localized food production and maintaining food security. The articles call for regaining control of the means of subsistence to produce food and satisfy other needs.

This strategy is criticized by two streams. Leftist libertarians appreciate the anti-globalization tendencies of this approach. However, they criticize it of being idealistic and anarchist. Trying to reform the current system through subsistence activities will be merely termed lifestye decisions in existing industrialized countries. These are likely to remain marginalized. Similar activities are likely to be phased out in the pre-modern South under international market economy pressure.

The second stream of criticism is headed by feminists like Bina Agarwal who question the romanticized idea of the "renaissance of earlier susbsistence relations based on traditional egalitarian political and economic mechanism". According to her, these practices never had any place for women's equality. Agarwal writes that attempts to revive subsistence economies could further entrench women in unremitting, undervalued labour. The main challenge today is to change the status of women from victims of environmental degradation to effective agents for environmental regeneration. This involves confronting social, class and political barriers (cited in Cochrane, 2009: 178–188).

Resonance with Bioregionalism

Regina Cochrane responds to Agarwal in her article, "The Sustainability of Tradition? From the Subsistence Perspective to a Feminist Ecoregionalism" (Cochrane, http//www.arts.ualberta.ca/cms/cochrane.pdf) Cochrane attempts to understand whether the ethos of the subsistence perspective resonates with "bioregionalism". She uses Sale's arguments to advocate human scale institutions, which are readily understood by people. In such institutions, individuals can actively cooperate with the people they directly impact. Sale's vision fixes the optimal extent of cultures, politics and economies to ensure that their interaction with nature remains sustainable and harmonious. Under such circumstances, the political systems of these communities need not conform to any norm – democratic or otherwise. She cites Purchase (1994), who tried to rework the concept of localized communities into an anarcho-syndicalist eco-regionalism. He elucidated a detailed structure in which workers' syndicates would organize industrial production of self-determining and humanly scaled city-regions. Thus, their cultures and economies would be integrated with local ecosystems. On the agricultural level, a small-scale polyculture will allow city-regions to become self-sufficient in food production.

Cochrane states that the aforementioned responses to environmental decline and globalized capitalism appear more adequate than a retreat to premodern traditions. She cites the failure of fisheries in Newfoundland (Canada) as an illustrative example. Both men and women were traumatized when they sought premodern explanations (through the Christian faith) of their loss. She avers that women are likely to lose in such situations.

New Economics Movement

Rajni Bakshi (2007) introduces the New Economics Movement by citing Edgar Cahn's (a community currency pioneer in the USA) quote which states: "There is more to the economy than the market. The earth is not flat. We need a new map of the real world. The map provided by government and prepared by economists is fatally incomplete. It defines reality exclusively in money transactions. That's all the GDP measures. It is a flat earth view."

The flat earth view is responsible for many wrongdoings of the modern industrialized world. It destroys the very ecosystem on which it subsists. It transfers enormous material wealth from the poor to the rich. The flat earth view is rooted in the mechanistic and reductionist approach of the modern scientific paradigm. New economics attempts to reframe many axioms of the industrial economy, including notions of efficiency and optimization. It is committed to the construction of democratic and institutional frameworks (of mutuality and cooperation) and rights and responsibilities. It also seeks the impartial enforcement of these mechanisms. New economics is flourishing in the space freed by the withering of the delusion that the world must choose between communism and capitalism (Bakshi, 2007: 43–44).

New economics does not claim to be value free or objective. It makes humanity face the ecological limits of its activities by turning the consumer society into a conserver society. New economics is weary of any moves that might lead to a centralizing, non-participative and potentially authoritarian state. It embraces the conscience of individuals and their links to the natural world without unrealistically idolizing these traits. Sustainability should come from limits internalized in the social fabric. It should not be imposed by force or by any group against another.

Paul Ekins, in his book, *Wealth Beyond Measure* puts forth concepts like priceless wealth. He also envisions a sustainable and universal abundance. Other concepts include as cited in ibid.: 45):

Wants and Needs

Green economics makes a clear distinction between wants and needs. It seeks to satisfy both. The focus is on welfare creation rather than satisfying insatiable wants. Elgin advocates 'voluntary simplicity', i.e., frugal consumption. People can change their lives by embracing voluntary simplicity, ecological awareness and personal growth. In the process, they have the power to change the world (ibid.: 46).

Wealth Distribution

There are two central objectives: the elimination of poverty and the maintenance of economy at an optimal ecological size (a level which can be handled by the biosphere). Notions of 'optimal' involves many aspects including the number of people, the ecological footprint of their consumption patterns, the ethical basis for extracting and using natural resources etc. This notion also questions whether our species has the right to reorder the rest of nature purely for its own convenience (ibid.: 46).

A Progressive Market and the State

The market and the state are seen in complementary roles. The market is required for the people to make choices. Green economics visualizes a progressive market in which consumers' choices include ethical, social and environmental considerations.

The state is required to make and enforce rules and to correct market failures. It has a key role in ensuring that polluters pay damages, monopolies are not formed, companies divulge requisite information to the public realm, basic human rights are respected, high quality education and health care are available to all and so on. The state should guarantee a decent, minimum standard of living to those who (through age or disability) have special needs.

Money

Green economics values money in its conventional functions. However, it also seeks various monetary reforms. Money should be a useful tool but should not be confused with wealth. The flow of money between companies, households and government is given due importance. There is also a need to take stock of 'social economy' (also called 'core economy' or 'love economy') which does not use money. According to an estimate, households and other forms of voluntary activity (self-help, barter etc.) produce the equivalent of almost one-third of the world's gross annual product.

Green money differentiates between wealth creation and money creation. Wealth is created by application of human skills to natural resources while money is a mere human contrivance. It is a symbol created by a deliberate process involving entities called banks. Hence green money imposes strict financial discipline on governments and controls consumer credit to guard against inflation (ibid.: 48).

Ecological Accounting

New economics seeks to displace Gross National Product (GNP) as the sole measure of prosperity. In the early 1990s, Paul Ekins and others proposed that various elements be added to and subtracted from the GNP to produce a more realistic indicator of real income. This could be called the Adjusted National Product (ANP). Ecological accounting identifies and monitors the crucial economic functions of environmental 'services'. New economics thinking was supported by the Brundtland Report's recommendation that there should be an annual report and audit to track environmental resource changes. Over the last fifteen years, various environmental and social well-being indexes have mushroomed. For example, the MDP (Measure of Domestic Progress) adjusts the conventional economic measure of GDP by subtracting the cost of crime, pollution and environmental degradation (as negatively affect the quality of life). MDP calculations show that social progress has been decoupled from economic growth in Britain since the last thirty years (ibid.: 50).

Proper Valuing of the Household Sector

Even in industrial countries, household production contributes 25–40 per cent of the GNP. It contributes an even higher percentage in less industrialized countries. According to Ekins, "There is no way that the wealth-creating contributions of these diverse groups can be given a financial tag. Yet, for many people, they are already the difference between life and death. And because they are tackling the big issues of our time – peace, human rights, environment and development – it is possible that they will perform the same service for humanity as a whole."

However, Bakshi quotes Ekins in stating that "the New Economy cannot be crafted in academic cloisters. The key moving energy comes from all of us .... In building a Green economy, motivation, information and moral commitments are more important than money."

Biomass-based Strategy: A Concrete Alternative, based on the Grassroot Initiatives

This section is based on K.R. Datye's book *Banking on Biomass* (1997) assisted by Suhas Paranjpe and K.J. Joy. The aforementioned strategies and those of other environmental philosophers are important for directing our future. Most of these visions are characterized by the decentralization of production bases, which ensures the decentralization of political power. The market, technology etc. are expected to fall in line. These visions also entail building new institutions.

These concerns are being logically and concretely built in the books mentioned above. The writings suggest where to begin, what steps are to be visualized, what contradictions are likely to emerge and their subversion. They also emphasize the necessity of building organizational bases among the marginalized to operationalize new ideas of resource use and new technologies. This exposition is premised on the linkages between productivity, sustainability and equity.

This basic premise provides environmental literacy and the eco-technical parameters discussed by Vandana Shiva. We need to identify the policies which will help us to decentralize production bases. This struggle should not defer finding the right strategy while advocating immediate concrete programmes for action.

Environmental movements in developed countries argue that such countries utilize far more energy than is needed. Further, most of their production is discarded eventually. The environmental costs of this extractive, energy-intensive economic growth are destroying the conditions of production itself. Hence, many ecologists and green activists are demanding a reduction in growth rates or even a zerogrowth economy.

The zero growth plan cannot be applied to Third World countries because the governments cannot ignore the livelihood needs of the impoverished majority. These countries have lesser access to fossil fuel energy than the North. It is neither possible nor desirable to colonize the resources of the non-industrial sectors. Moreover, their resource base is directly affected by the energy use pattern of the North.

Datye claims that Third World countries (particularly India) must enhance productivity *without compromising sustainability*. This will be possible only if they make bio-resources their main energy supply. The bio-resource base should be utilized to increase the efficiency of energy use. Third World countries have favourable conditionalities, like a massive human labour resource base, contemporary technological advances etc., to realize this paradigm.

Environmental Blindspots

Datye notes that organic farming advocated by environmentalists for advanced countries is inadequate. We need to ascertain whether the use of external inputs (chemical fertilizers, pesticides, water etc.) can be optimized to the extent where their short-term gains do not impair the productivity of the ecosystem but maintain or enhance it.

These processes require a high degree of proficiency in bio-energy production and resource management. Creation of physical assets should concur with human resource development. The current skills of the entire workforce and the management capabilities of groups and communities need to be identified and enhanced. Datye states that considerable skill is required to isolate those who appropriate developmental gains and to curb the power of free riders. There should be an equitable distribution of the benefits of enhanced productivity.

Datye insists that in a small farmer-artisan economy, an increased level of productivity and income generation depends on the adequate and assured availability of essential physical infrastructure facilities. To grow beyond subsistence, this mode requires small but definite quantities of external inputs, supportive inputs for skill upgradation and supportive structure for value-added processing. Datye does not glorify traditional technologies because the energy and commodity exchange between (a) urban society and the industrial sector, and (b) rural communities and agriculture (along with other allied sectors) is essential for increasing the carrying capacity. The dogma of 'self-sufficiency' (characteristic of traditional societies) is to be avoided. The goal should be self-reliance and a balance in this exchange.

Datye also articulate the important economic principles of this process:

- Distress sales should be avoided.
- Food security and income stability should be assured to the resource poor.
- The local energy resource and surpluses should be generated in the long run. Value added processing in dispersed industries (based on energy surplus) should create a sizeable surplus of processed commodities.
- Marketed surplus of commodities and energy at each stage of development should match external inputs.

Once the unequal exchange of energy flows is controlled, the standard of living of the local population will increase thereby eliminating their exploitation. It is crucial for those who accept this strategy to understand the dynamics of energy flow. The dependency school had discussed unequal exchange when they had proposed the import substitution strategy in 1960s and 70s. However, Immanuel Wallerstein (1974) argues that capitalism has developed as a world capitalist system since its inception. The production processes of Third World countries are linked to this process through colonization. Datye points out that avoidance of unequal exchange is a serious issue in the world capitalist system. However, one must acknowledge that economists and political scientists have not viewed the exchange as mainly taking place between energy flows. They do not accept that this flow is also controlled by political economy (rather than just political power). The emergence of ecological concerns has been accompanied by an increased development of the science of energy flows. People are more aware of **unequal energy exchange**.

Energy Processes: A Change in Paradigm

Ecologists now view the planet itself as a living organism which has been functioning for millions of years by constantly increasing its order and complexity. Datye quotes Odum (1991) in stating that the study of the ecological food chain indicates that natural ecosystems and other natural self-organized systems develop designs and pathways which do not simply dissipate but conserve and reinforce energy use. In fact, this is a precondition for their emergence and change. Datye insists that we should see humans as part of this process. We need to evolve methods of organization which optimize the use of external inputs to enhance basic planetary processes (which are the basis of all life).

Datye again quotes Odum in saying that magnitude changes every time energy is transformed from one form to another. Human energy is of a higher order than the bio-energy of plants and animals while bio-energy is of higher order than solar energy. The incorporation of human beings in this scheme adds a new kind of order into the system. An added dimension of human energy is information and its use for conscious regulated activity.

Datye elaborate two characteristics of the energy scenario in nature. Firstly, a so-called hierarchy in natural systems is an important regulatory node (of the web of relations) which ensures the self-reproduction of the whole. In natural systems, a single unit of energy of a higher order represents about 1,000 times the unit of lower order energy. Increase in one unit of higher order energy represents an expansion of the self-reproducing net.[16] Contemporary human systems are not organized in this way. Within human systems, large masses of energy cross over to the energy sink without being involved in the self-reproductive processes. This eliminates sustainability.

Secondly, the hierarchy in human society is a purely social construct and does not have an ecological functional basis. It depends upon social arrangements which tie access to energy and resources to one's location in that hierarchy. Datye suggests that in human systems, so-called higher levels (state, administration) should act as nodes which expand access to resources to more people. They should ensure that production is based on local and renewable resources. Services should enhance productivity and efficiency in the use of scarce resources.

Productivity, Sustainability and Equity

The concept of productivity discussed here refers to primary productivity, i.e., the productivity an ecosystem would have if external inputs were withdrawn from it. Primary productivity is measured in terms of biomass or (strictly speaking) the primary glucose synthesis potential of an ecosystem. This concept of productivity is integrated with the concept of sustainability. Current systems of high input agriculture decrease primary productivity. The apparent, visible productivity is based on secondary productivity which depends on external inputs.

Datye reveals an important political stand. This perspective does not conflict with an equitable, radical redistribution of present productive assets (as suggested by leftist politicians). However, it does not treat redistribution as a necessary starting point. Mere redistribution (e.g., land in West Bengal) does not create sustainability in the absence of sensitivity and availability of associated inputs. They feel that there is enough space within the current ownership pattern to redefine equity and ensure livelihood on the basis of a much wider social consensus.

Equity is seen here in terms of the right to equitable access to and collective participatory control of a certain class of assets and resources. These are resources which need to be dissociated and treated independently of current productive assets. This class includes resources which are common property and renewable energy resources generated through social action and infrastructure facilities like watershed development programmes. Common property resources should be widened to include water, waste and infrastructure.

Strategy in Brief

This eco-development perspective concentrates on the realization of food security, fuel and fodder self-sufficiency. The long-term goal is to create a 'bio-energy' surplus above susbsistence needs within a reasonably short time horizon so that one can pay labour charges to develop assets for biomass regeneration under employment guarantee schemes. This bio-energy surplus represents a renewable energy resource in terms of fuel and materials. It provides input for the rapid development of infrastructure and later (in combination with renewable energy development) a base for the transition to a dispersed, modern but sustainable industrial system.

Datye has concretely charted out his suggestions for efficient bio-energy production. They also list potentially supportive policy formulations. The state has to play a nodal role in establishing some kind of a regulatory regime. They are well aware that the role of the state is receding within a global regime. However, poor people would like to see the state persist as a relatively autonomous block compared to any specific industrial group or TNC. The democratic state owes this to poor people, whose votes give it the mandate to exist.

Datye's strategy for technology has two main components: (a) To make water available for wider land use and providing minimum quantity of water to everybody including landless labour who can lease the land and use the water to satisfy livelihood needs. For this, they suggest the in-conjunctive use of water impounded in big dams along with locally harvested water (through watershed development programmes etc.). (b) Energy needs should be similarly satisfied by the in-conjunctive use of fossil energy and locally created solar, wind and bio-energy. The authors do not preclude the conventional industrial method. However, they advocate its judicious use along with biomass based raw materials and other energy options.

Their strategy proposes decentralization to enable access to natural resources. They advocate the equitable distribution of this access to ensure that production bases remain dispersed. This will encourage the decentralization of political power. Women should have separate access to resources. They should not be subsumed within the family enterprise.

Their ideological strategy appears to be based on what they call advanced science or what Carolyn Merchant has termed 'postmodern science'. This science believes in strengthening ecosystems and maintaining primary productivity so that external inputs are used in an optimal way. It develops individual knowledge and skills to a high degree. Such knowledge is gleaned in the absence of simple mechanistic quantities of external inputs (to be administered to agriculture). This absence makes it necessary to monitor and steer natural processes. This is a utilitarian view of nature as Datye's primary concern is to abolish the poverty of marginalized people in the Third World. However, in their strategy formulation, they are aware of the natural processes of interactions and cyclical flows. Thus, they foreground the need to work in harmony with nature.

Datye has also tried to analyse the relevance of the Gandhian vision of self-sufficient villages to their own proposed vision. They appreciate the Gandhian concern for full employment and livelihood security. They note Gandhi's insistence on the dignity of labour and equality. They criticize Gandhi of overlooking the caste barrier which is hurdle to creating learning opportunities for artisans, farmers and women. The debate between Ambedkar and Gandhi becomes important in this light. Ambedkar saw industrialization and urbanization as necessary and felt a need for centralized production processes. According to him, that was the only way to assimilate dalits into the mainstream and negate the caste system. However, centralization creates wage labour, i.e., the proletariat. Mies and her colleagues oppose the production process based on 'alienated labour'. Datye visualizes that his idea of a new paradigm and dispersed industrialization processes will negate cast barriers and iniquitous distribution of natural resources. It will create a community-based self-reliant production system.

Datye reminds us that Bhoodan (a Gandhian initiative after independence) did not succeed because of politicians and the bureaucracy. Women were left out of land reforms. No infrastructure, technology, special credit or market facility were created. It is time to revisit the Gandhian precept of "unto the last". Gandhi's vision was firmly based on rural areas which is unacceptable to the modernization paradigm. The social and ecological costs of maintaining such a system of production have been externalized in Western social democracies by the global acquisition of territory and energy resources. A production system based on energy self-reliance can productively utilize the surplus mobilized by rural, small town and urban communities. Thus, they claim that this alternative strategy has a critical but fruitful relationship with Gandhi's thought.

While developing an idea of unequal exchange, Datye could also have given the example of women's energy spent in reproduction such as daily chores (which is not returned in equivalent units). Women's energy (expended in reproductive work) acts as a node for self-reproduction and contributes to an expansion of the complex web of human life. However, their basic understanding of the ecological processes is convincing. Women can work out a better position for themselves within their vision.

11

Criticism against Ecofeminism and Response

> Is ecofeminism a social movement to empower the dispossessed? Some major themes in ecocentric and ecofeminist discourses are the rejection of self-determination, scienticism and instrumentalism; the promotion of sexual essentialism; myth making and the prescription of subsistence utopias. (Cecile Jackson, 1995)

This chapter examines the criticism levelled against the ecofeminist perspective by Marxist and liberal feminists. This scathing attack has ousted ecofeminism (at least the version of Vandana Shiva, Maria Mies and her co-authors) from mainstream feminist discourse. Ecofeminism has been alienated from development policies which are pursued to change the condition of poor women, especially in the South. Carolyn Merchant's writings have not come under criticism. This can be attributed to the fact that they do not imply political action against the present processes of the world capitalist system. Her writings also do not suggest strategies for future development efforts. This critique appeared in the reviews of the book *Ecofeminism*, jointly written by Mies and Shiva. The reviews included references to their other writings, particularly Mies' *Patriarchy and Accumulation on a World Scale* (1986) and Shiva's *Staying Alive* (1988). *There is an Alternative* (2001) edited by Veronika Bennholdt-Thomsen et al. was also mentioned.

Points of Criticism

Except the partisan review of Rhonda Roland Shearer (*Signs*, Winter, 1997), Maxine Molyneux and Deborah Lynn Steinberg (*Feminist Review*, Spring, 1995), Cecil Jackson (*New Left Review*, 1995, No. 210), R. Braidotti et al., in their book, *Women, the Environment and Sustainable Development: Towards a Theoretical Synthesis* (1994), Govind Kelkar and Gail Omvedt (an unpublished paper: 1993), Naila Kabeer, in her book *Reversed Reality* (1994) have derided the ecofeminist stance as universalist, essentialist and unrealistic. Their criticisms are listed in brief as follows:

Firstly, there appears to be a one-to-one conceptual connection between patriarchy, scientific methodology, markets and ecological destruction. Women are depicted as direct victims of this power block and as having no agency of their own.

Secondly, ecofeminism is accused of methodological confusion. Reviewers cite that while Mies and Shiva blame scientific methodology of using dualistic rationality and dichotomous pairs like nature/culture, women/men etc., they themselves are using the same understanding to uphold the other extreme. Mies and Shiva universalize categories while criticizing modern science and its epistemology.

The reviewers are disturbed by Mies and Shiva emphasizing the connection between the power of modern science (as a monolithic institution) and its identification with the male. The reviewers argue that their analysis is an absolute notion of male identity and power. The language used by Mies and Shiva supports biologistic determinism. Even though Mies and Shiva criticize science of adopting the reductionist approach, they themselves construct their criticism using the similar totalizing and reductionist paradigm.

Naila Kabeer states that in Mies' approach, there is an overriding emphasis on the Big White Men and their power which epitomizes women's exploitation and oppression (Kabeer, 1994: 52). Mies does not disaggregate these categories by questioning the extent of their coherence.

Shiva is criticized of generalizing the origin of the oppression of women and nature. She avers that this oppression began with Western scientific imperialism and is a modern invention. Shiva's focus is the comparison of mechanistic versus organic thinking. Her examples include the medicalization of childbirth (with the corresponding fragmentation and fetishization of the female body) and biogenetic technologies (which convert nature into a genetic resource to be engineered, patented and owned for corporate profit).

Another contentious issue is the romanticization of the female, particularly the indigenous Third World female. The standpoint which originates from such a location is called the 'view from below'. It is frequently romanticized as producing real truth. The 'view from below' is contrasted to the 'view from above' (in this case the scientific point of view) which by professing to be objective and 'value free' produces false truth. This is cited as an example of essentializing the Third World women.

Thirdly, although Mies and Shiva insist that ecofeminism constitutes a materialistic critique and politics of capitalist accumulation, patriarchal violence, the oppression of women and the corresponding resistance, neither of them defines or traces these issues as specific processes. This essentializes capital, cash and patriarchy as disembodied/abstract agents with will, motive, feelings and drive. A refusal to be specific or to acknowledge partialities ultimately denies that science is socially constructed. It also eliminates any chance of feminists being able to change it. Thus, their method appears to be counter-productive to their own agenda.

Fourthly, the arguments presented by Mies and Shiva about the "postmodern critique of universalist (Western) ideologies" appear contradictory to the reviewer. According to Mies and Shiva, the postmodern critique has failed. They posit that the cultural relativism of postmodernist feminism is a result of the failure of universalist ideologies (Marxism, socialism etc.). Both Mies and Shiva explain this failure as the 'suspension of value judgment'. The reviewers feel that the authors' understanding of this failure is confusing. They assert that this confusion occurs because Mies and Shiva do not understand power relations. This lack of understanding is embodied in their vision (which itself is based on alternative universalism). This alternative emerges from the "allegedly intrinsic commonalities of grassroots ecofeminist activism".

Fifthly, a lot of criticism is directed against upholding the metaphor of 'Mother Earth' for nature. This metaphor is based on identifying women with their reproductive ability and will for maternity. This translates into accepting the image of woman mainly as a mother. It also accepts heterosexuality as the ideal way of life. Reviewers state that "Mies's description of sexual intercourse as (ideally) a celebration of nature seems to posit ecofeminism as a standpoint, which is intrinsically and necessarily heterosexual" (Molyneux and Steinberg, 1995: 101). It carries the implication that rescuing heterosexuality is central to the rescue of 'nature' and 'women'. The writings of Shiva and Mies are characterized by a striking absence of fertility control or women's capacity to exercise choice. In their writings, this choice is often contrasted with reappropriating the traditional knowledges of women. The reviewers object to the glorification of traditional societies as they believe that women in these societies did not have a wide array of choices.

The reviewers have also charged Mies and Shiva of making two significant assumptions about the character of ontology and epistemology: (a) ontology is essential rather than contingent (there are intrinsic commonalities in women's experiences which override their differences), and (b) ontology equals epistemology (the knowledge of women emerges from/equates with the being of women). The reviewers assert that the problem of avoiding cultural relativism can be better addressed by analysing power relations. A basis for unity and resistance is not found only in a presumed ontological commonality (women as victims, or colonized). A shared political agenda (developed through a range of situational standpoints located in specific and complex material relations of space, place and power) can also help achieve commonality. Bina Agarwal (1992) agrees with this criticism.

Criticism by Cecile Jackson, Govind Kelkar and Gail Omvedt, and Braidotti et al.

Jackson has abstracted (from Mies' proposal) that self-interest must be overcome to develop women's solidarity internationally: "Because the core of individual freedom, equality and self-determination (linked to money and property) is the self-interest of the individual and not altruism or solidarity, these interests will always compete with the self interests of others" (cited by Jackson, 1995: 127). Jackson comments that Mies' arguments unnecessarily dichotomize certain definitions of altruism and self-interest (while overemphasizing the significance of Western women to international solidarity). She accuses Mies' working assumption that the Western individual is as stereotypical as her version of 'Third World women'.

Jackson also analyses the idea of the 'subsistence perspective'. She states the prescription of this perspective by Mies and Shiva is the product of "the search for an ecologically sound, non-exploitative, just, non-patriarchal self-sustaining society characterized by disengagement from international capitalism, regional food self-sufficiency, basic needs for all, development from below and the end of anti-rural bias. Women play a key role ... in grassroots movements ... it is women who understand that a subsistence perspective is the only guarantee of survival for all."

Jackson comments that self-sufficient rural livelihoods exhibit high levels of unpaid women's work. The moral economy does not guarantee equal survival for all (women in such societies frequently have reduced access to food and health care than men). 'Development from below' unrealistically assumes gender-neutral village-level institutions. Jackson asserts that non-monetized subsistence relations can be deeply exploitative. She claims that involvement with markets has often been progressive for women. Govind Kelkar and Gail Omvedt cite examples of women farmers who want to produce for the world market and the trade union women in South East Asia (who have welcomed the shift of jobs from women of the North to women of the South).

R. Braidotti questions whether subsistence agriculture will be able to feed the densely populated areas of Europe and India (Braidotti et al., 1994). Braidotti reiterates the criticism of Indian Marxist feminist scholars against Shiva in claiming that she idealizes Indian subsistence agriculture and recasts a past where people lived in perfect harmony with nature. She supposes that women were highly respected in such a society. Thus, Shiva is accused of ignoring caste and patriarchal tensions. By glossing over the exploitative structures of Indian society, she blames the environmental crisis wholly on the 'state' and the global economy.

Govind Kelkar and Gail Omvedt (1993) focus their criticism of Mies and Shiva on three points:

1. Western bankers push peasants in the Third World to produce for the world market. They pressurize these peasants to withdraw their labour from providing their own subsistence needs. Mies and Shiva reject their entrance into the market.

2. The technologies promoted to produce food (and some other products) are rooted in markets and hence are inherently accompanied by violence and destruction. They need to be rejected.

3. According to Mies' criticism, the strategies (of integration) pursued by one section of the feminist movement to 'catch up with men' like the policy of equalization, positive discrimination and special quotas for women indicate that women want to share the model of 'privileges of victors'. Hence, these should be rejected.

My Reading of Mies, Shiva and Mies' Co-authors

While documenting the methodological flaws of Mies and Shiva (and expecting rigorous and substantive arguments), one must remember that both of them accept certain ecological premises. Their perspective has been developed from this reality. Both believe that there are limits to the use of natural resources. They also believe that markets will not benefit the poor in the Third World as the conditionalities which existed in the West (in the beginning of the expansion of capitalism) do not exist anymore. Hence, there has to be a different perspective for development contrary to the Marxist perspective (which still upholds the development of productive forces to emancipate humans and particularly women). Against this background of contradictory perspectives, every statement expounded by Mies and Shiva within the ecofeminist framework is found flawed or methodologically incorrect. Both are universalists who share their value judgements of equality with Marxists and socialists. The additional categories (men, women, non-human nature etc.) contribute to their proposed strategies under the 'subsistence perspective'. These strategies appear contrary to the Marxist and socialist strategies of 'development of productive forces (technology) and instrumentalization of nature'. Marxists have also propagated the 'proletarian view' as having libertarian potential (i.e., they support the 'view from below' for the emancipation of humankind). Mies replaces this category by 'women of the Third World' who are involved in subsistence farming and providing basic food security to men and children. These women are the reserved army pool of labour for the world capitalist system. Mies asserts that unless subsistence needs are made central to the development process, inequalities between men/women, urban/rural communities and developed/underdeveloped nations cannot be reduced. Without foregrounding these values, surplus accumulation processes cannot be arrested. According to her, modernism makes individuals selfish and atomized (due to market processes). It propagates the illusion of choice and self-determination while escalating inequality between nations and within national boundaries. This is a battle between two diametrically opposite perspectives based on different premises.

Critics seem to be forgetting that theoretical analysis is conducted at various levels what Marx calls 'determinate abstractions'. Mies and Shiva deal with the highest level of abstraction. They are building a multidimensional world view, i.e., a conceptual framework. This exercise is partially totalizing. However, it is being done on the basis of diverse historical materials (discovered since the second wave of women's movement began). Mies has reinterpreted this historical material. Her work includes references to primary published studies (of investigative nature) and experiential testimonies.

Mies' exercise is in the genre of Marxist ideologues who abstract from already published material. Using this as reference, they build a new world view or paradigm to analyse the constituents of reality. Mies' analysis has incorporated 'nature' (ecology) and 'gender' in this reality. She views women's reproductive work as the basis of male productivity and the reproduction of society. From this base, she suggests strategies for transformation and change. Even Marx had developed a materialist, dialectical conception of history (in *German Ideology*) to analyse and explain the transformations of a dynamic, ever-changing world. Many still accept (with some revisions) his methodology and basic conceptual formulations to understand contemporary situations.

The premise of Mies and Shiva's perspective is constituted by two major conceptualizations. The first is an understanding of the historical-materialist and the dialectical conception (of women, men, humans and non-human nature) where the interaction of these elements provides dynamism to changing reality. It is incorrect to call Mies ahistorical because (after exploring the origins of patriarchy) she postulates that patriarchy is an evolving category. Along with class exploitation and accumulation processes, patriarchy reinforces the hierarchies of all societal institutions. It loses its private character in capitalism (to some extent) but is reinforced in the public sphere. The second conceptualization is related to the colonial process of capital accumulation which nurtures the world capitalist system. Thus women are stratified at the global level. The First World women's privileges are possible because of their share in the surplus accumulation process. The Third World subsistence farming women remain at the bottom of this hierarchy. These conceptualizations help us to understand the conflict between the two perspectives.

While criticizing Mies' writings, none of the aforementioned feminist critiques introspect into their own premises and modernistic world view. This view was adopted before ecological movements affected mainstream thinking about development processes and sustainability issues. The First World feminists do not analyse the colonizing implications of the world capitalist system. This system has co-opted them into the hierarchy of exploiters and posited them against Third World women (whose cheap labour contributes to their rich lifestyles). Their supremacy is achieved at the cost of the health and natural resources of the Third World. Competition on a global scale leads to reckless productive practices. Mies wants feminists to engage in efforts to disentangle themselves from this chain of exploitation and accumulation. According to her, this chain reinforces and expands the inherently structured inequality between nations and within national boundaries. She criticizes mainstream feminists of ignoring the destructive tendencies of capitalism. Mies avers that these feminists merely indulge in market-based capitalist ideas of the sustainability of the existing economic system (by improving market access and tinkering with so-called development policies).

Food security influences the central theme of her argument against technology. She states that certain kinds of technologies disturb the ecological base and are life threatening. Markets allow the internalization of depletion costs by discounting and compromising natural resources for future generations. Several fellow environmentalists share the same vision.

Mies' passion for decentralized societies and community living is severely criticized by feminist critics who cling to modern values of self-determination, scienticism and instrumentalism. These critiques invariably lump these values with sexual choices, fertility control etc. They also fear that an emphasis on 'mothering', 'nurturing', 'sexual intercourse as source of pleasure' and the likening of the 'regenerative' power of women to that of nature is a denial of modern life. This might relegate women to the days of feudal family labour and the mothering of children (to provide for farming). Mies envisions completely new communities which are created on the basis of shared values and negotiations.

A useful example is that of the Dalit Graduate Group in Mumbai. This group believes that although Ambedkar encouraged Dalits to leave their traditional lowly duties in the villages and migrate to cities (to utilize opportunities in a casteless atmosphere), they lead precarious lives in city slums. This has a criminalizing influence. The Dalit Graduate Group advocates that Dalits should imbibe education and assert themselves in village life. They should take advantage of the different schemes meant for them. Once certain new values are imbibed, they are not easily flushed from the system. Mies and others do not romanticize old communities. They are searching for an alternative to centralized authority (both the socialist state and the globalized capitalist economy). The images of mothering and nurturing symbolize the need to regenerate society and preserve natural bases. Mies asserts that women (especially the Third World women who struggle to survive and regenerate) are best suited to this task. They can teach us to use resources frugally. This attitude is essential because there is a limit to growth.

Mies envisions an overarching equality in all continents. She does not limit herself to equitability (which depends on the individual's specific location in society and is related to geography, class, caste, gender etc.). In their universalistic framework they appreciate that women (in the beginning of the women's movement) discovered commonalities in terms of their exploitation and oppression within the capitalist patriarchal system. The fight against sexism and violence was common to all women in all countries. They are grieved by the current women's movement which substitutes liberation with empowerment (where women want money, freedom and equality). They are critical of the discourse of Northern women who (by virtue of the privileges of White men) ignore that they are priviliged vis-à-vis women and men of the South. This may be the reason why they shy away from working out a better society (Mies et al., 1999: 206).

Mies' views are criticized on the grounds of ignoring the situational nature (located in material relations of space, place and power) of political agenda. Mies and Shiva appreciate the epistemological challenges posed by postmodern feminists to knowledge construction. Postmodernists criticize the universalizing rationality of modernist discourse. Mies avers that postmodernists have employed the cultural relativist approach in response to the failure of socialist societies. She asserts that promoting diversity is as faulty as the universalizing theories of socialism. According to her, this perspective encourages a limited micro-level and selfish agenda. It divides women into tiny pockets and then allows them to decide their definition of equality or the ideals of liberation. This eliminates the burden of colonial power and accumulated wealth from the shoulders of First World women. Thus, it negates the responsibilities of First World women (which are rooted in their historical advantages in the hierarchically structured world).

The 'romanticism' and 'idealism' of the strategies of Mies and her colleagues is illustrated by using successful 'subsistence communities' as illustrative examples. These communities thrive within the globalized capitalist patriarchal world order. They are presented under the title "There is an Alternative" to oppose the usual cynicism of so called democratic, progressive and well-meaning policy makers.

I agree that these examples do not present a clear road map for dismantling the present order. They do not illustrate how to disengage from the system and start a new society. These solutions sound anarchist precisely because they do not present solutions. Communities have to engage with the state to plan their own development. They need to learn how to procure funds from the state to gain autonomy. Decentralization should be a conscious policy to be implemented progressively.

Mies' views are valuable because of her clear focus and critique of the nexus of scientific methodology, technology, markets, colonial processes and ecological destruction. One can discern the roots of poverty, gender inequality and the need for appropriate technology (in the process of women's liberation) through her vision. She also helps us understand the larger processes of human liberation and their inevitable anchoring within the processes of ecological regeneration.

12

Illustrations of Sustainable Natural Resource Use Practices

> The Pani Panchayat follows certain pre-determined principles. The irrigation schemes are undertaken for groups of farmers rather than individuals. It believes that water is a common property resource and all villagers (irrespective of land ownership) should have an equal right and access to its utilization. Membership is offered to landless villagers to enable them to become sharecroppers to farmers having more land. They therefore gain employment in the village itself. Thus land rights are detached from land ownership. The absentee landlord is not given water. Water (made available for irrigation) is used for cultivating seasonal crops requiring less water rather than crops requiring perennial irrigation and large quantities of water specifically sugarcane. (Salunkhe, Lokur and Pangare, 1997)

Environmental movements fight against development policies which cause degradation and dislocation. They illustrate the sustainable and equitable use of natural resources to reduce poverty. Sustainability cannot be achieved without incorporating the equitability principle. It is difficult to glean support for alternative development models from governments (entrenched in the market economy) and development paradigm (antagonistic to zero-growth or non-monetary economics). Kuhn has revealed that paradigm shift takes time. A critical mass of practitioners and theoreticians is needed to develop any rupture. The problem lies in translating these visions into strategies for action. Such strategies will encourage critical masses of willing people to act (thereby supporting the environmental movement). It is difficult to restructure communities which share the ecosystem and willingly live according to its exigencies. This difficulty arises out of divisions based on gender, class, caste and racial inequalities. Some illustrations of good practices of natural resource use in India are presented in this chapter. Some of these projects have not been framed with a long-term vision. However, they assimilate some basic ecological and social principles as basis.

Sumi Krishna's book *Livelihood and Gender* (2004) presents researched documents on community rights within natural resource management. She focuses on the process of genderization in this process. She observes that the thrust towards greater state control of natural resources (since the early 1980s) has been countered by decentralized community resource management. Such a paradigm is likely to empower women. Increasingly reality is being proven consistent with vision. Krishna states that establishing linkages between praxis and theory (in gender and NRM) is difficult. High-profile approaches (notably ecofeminisms) have gained wide popularity without being implanted in rigorous empirical researches. Practitioners engaged in action lack the time, opportunity, inclination and training to theorise. Women's affinity to nature and their traditional knowledges of natural processes cannot be taken for granted. Krishna distinguishes between management and rights. She wishes to ascertain whether women's rights in NRM are premised on the sexual division of labour. She avers that emphasizing women's experiences within participatory NRM is not an essentialist tactic. Krishna's book reveals a nuanced understanding of women and their concern for ecological regeneration.

Forests

Forestry covers the second largest land use area in India (after agriculture) of about 22 per cent of the total land base (World Bank Report, 2006). Approximately, 275 million people in rural areas depend on forests for at least part of their livelihoods. This livelihood is derived from wood fuel, fodder and non-timber forest products like fruits, flowers and medicinal plants. About 70 per cent of India's rural population depends on wood fuel for domestic energy needs (ibid.: 2006). Forest dwellers, including tribals, are among the poorest and most vulnerable groups of society. The most disadvantaged sections of these groups live on the fringes of forests. They tend to have a close cultural and economic bond with these forests. Tribal communities and rural people depend on forests for several basic needs (like food, fuel, fodder, small timber for houses, agricultural implements etc.). Artisans use bamboo and other types of wood to make baskets, toys etc. A major portion of their cash income comes from the collection and sale of non-wood forest produce such as tendu leaves, grasses, Mahua, Sal, seeds etc. (Kulkarni, 1997: 337). Tribal communities' lack of control over the resources that sustain them causes poverty. Most tribal groups are landless or have unviable landholdings. Forests are the fall back option for poor villagers when the rains fail. Hence, forests are an integral part of agrarian economy as they provide fodder for livestock, manure for fertilizing fields etc. Forests are exploited by resident tribal and rural communities with ingenuity and skill to meet their livelihood needs. Political and economic marginalization makes it difficult for these groups to practise ecologically sustainable lifestyles. Most of them overexploit forests under the influence of poverty.

In India, deforestation began with British colonialism in the 19th century. Before British rule, forests were manned by local communities. There were customary rules to protect forests, especially sacred groves and sacred trees (ibid.). British administrators realized the commercial value of forests. Lord Dalhousie's memorandum in 1855 retained teak timber as state property. The Forest Department was established in 1856. In 1865, the Government Forest Act assumed control over forests. Further laws passed in 1878 and 1927 supported large-scale deforestation of timber. This was exported for building ships, railway sleepers etc. The National Commission on Agriculture (1976) advocated the commercialization of forests while disregarding the sustenance of local people. It stated that the free supply of forest produce to rural people and their rights and privileges have destroyed forests.

In the early 1980s, there was a strong shift towards forest conservation. The parliament adopted the new Forest Policy Resolution in December 1988. It stated that the primary aim of forest policy was to ensure environmental stability and the maintenance of ecological balance (including atmospheric equilibrium vital for the sustenance of all life forms). The life of tribals and other poor people (living in and around forests) depends on forests. Their rights and concessions should be fully protected. Some of its main features were:

- Complete protection of villagers' customary rights and concessions.
- They should have primary access to forest produce for their domestic requirements such as wood fuel, fodder, non-timber forest produce (NTFP), construction timber, etc.
- Enhancing their income and employment by improving and increasing production of NTFPs.
- A "massive people's movement with the involvement of women" should be created for achieving policy objectives (cited in Sarin, 1997: 267).

To translate these objectives into practice, the Ministry of Environment and Forests, Government of India issued a policy circular in 1990. This circular listed guidelines for the involvement of village communities and voluntary agencies in the regeneration of degraded forests. This was formally adopted as the Joint Forest Management (JFM) approach. JFM's primary focus is to protect and conserve forests and forest-dependent communities. It currently covers 27 states, 85,000 village JFM committees and 17.3 million hectares of forest land. Villages included in JFM programmes encompass 8.4 million families. Assuming an average family to have four people, this figure represents 34 million people. Tribal families account for one-third of all families in JFM programmes (World Bank Report, 2006).

Apart from state governments, state forest departments and forest-dependent communities, the other stakeholders of this approach are Self-Initiated Forest Protection Groups (SIFPGs)

(Sarin, 1997). SIFPGs symbolize grassroots community initiatives for regenerating degraded forests. They deal with adversities caused by resource depletion. SIFPGs represent a revival of community resource management traditions (which were eroded by state intervention). They protect thousands of acres of state-owned forests in Orissa, Bihar and several other states on a smaller scale (Mohanty, 1996; Sarin, 1994, 1995; Poffenberger, 1995; Vasundhara, 1996). West Bengal has the largest JFM programme with approximately 2,423 forest protection committees (FPCs) protecting 3,90,919 hectares of public forests.

JFM fosters partnerships between state forest departments and local institutions of forest users and right holders. This is achieved on the basis of sharing benefits and forest management responsibilities. The state JFMs grant participating villagers free access to specific non-timber forest products and a 25–50 per cent (net or gross) share of poles/timber (in cash or in kind) of 'final felling'. In return, the villagers protect forests by forming an organization conforming to the membership and structure prescribed by state forest departments. Most states prescribe a forest officer as the local institution's member-secretary. This person is responsible for conducting all proceedings. Local community institutions are more effective in protecting forests than state forest departments. Recently, the results of a study conducted in Uttarakhand were published in the proceedings of the US National Academy of Sciences. The study compared satellite images of forests separately managed by village councils and the state. The study concluded that the level of degradation was similar. However, this needs to be analysed in the light of the fact that the Uttarakhand government spends seven to nine times more money on conservation than local communities.

Gender and Development in the National Context

According to Sarin (1997), till the early 1970s, Indian development planning was governed by the assumption that the household was a unit of similar interests. It was also assumed that the benefits of available resources are shared equitably (irrespective of gender). As a consequence, it was presumed that women's needs were satisfied. "Towards Equality" (a report of the Committee on the Status of Women in India, CSWI) (GoI, 1974) altered the government's perception of women's role in development. The report revealed that the dynamics of 'development' had created new imbalances and disparities which disadvantaged women. These problems included a declining sex ratio, reduced life expectancy, higher infant and maternal mortality rates, declining work participation, illiteracy and rising migration (GoI, 1988a). The report recognized that women were adversely affected by the processes of economic transformation.

"Shramshakti" (GoI, 1988b) – a report of the National Commission on Self-Employed Women and Women in the Informal Sector – reviewed the appalling poverty and exploitation of poor rural women who were struggling to support their households. These women did not own any productive assets. Hence, many of them resorted to collecting NTFPs when no other employment was available.

Women's lack of access to and ownership of productive resources has been identified as an important cause of their vulnerability. The Government of India took focused initiatives during the 1980s to counter the constraints on women's development. These initiatives are embodied in the Sixth Five-Year Plan (1980–85) (through joint titles to spouses during distribution of land and home sites) and the Eighth Five-Year Plan (1992–97) (in which the state governments were to allot 40 per cent surplus land to women alone and the rest jointly to spouses) (GoI, 1992). Programmes like JFM provide independent access and entitlement to forest resources to women. Therefore, they support resource-poor women.

Inadequacies

Although JFM represents 'participatory' forest management, its primary management objective is predefined and non-negotiable in almost all state JFM orders. The villagers' concerns are often ignored. The 'benefit' is a percentage share of the income from timber on 'final' felling. This allows village men and women to protect only degraded forests. Local institutions (as a whole or its individual members) are entitled to share the 'major' benefit only after 10 to 20 years. People who currently depend on the forest (for head loading, grazing, firewood collection etc.) have to forsake their consumption (Sarin, 1997).

When restricted from collecting firewood, women and men are compelled to either 'steal' firewood from closed areas or to travel extensively for collection. The former reinforces the stereotype of women and marginalized groups being forest destroyers while the latter increases their vulnerability to harassment and humiliation by outsiders. Such enactments increase the labour and time required for gathering cooking fuel (AKRSPI, 1995; Sarin, 1994, 1996; Sarthi, 1994; CES, 1995). It is assumed that access to a few low value NTFPs (under JFM) is adequate for supporting people's 'needs' (for firewood, fodder, small timber and other NTFPs mandated by the 1988 forest policy). However, many poor men and women derive their primary livelihood by processing or selling a diverse variety of NTFPs.

Land

Arable land is a critical determinant of economic well-being, social status and political power. In South Asia, most arable land is privatized. In India 86 per cent land is privately owned. About 89 per cent rural households own land albeit in small plots. Land ownership is crucial to those rural poor who depend on agriculture. A key factor to rural poverty is access to land. Land is important economically because it is a source of direct production. It can act as a collateral for credit (as a mortgageable or saleable asset during crisis). Land improves bargaining leverage with employers. Further, land owning widows and elderly are better cared for by kin. Arable land is the most valued form of property for its economic, political and symbolic significance. It is a wealth creating and livelihood-sustaining asset (Agarwal, 1997).

Land titles are important for women because of the growing feminization of agriculture (as men shift to non-farm employment). It enlarges their information base of indigenous seed selection and cultivation methods. In female-headed rural households, land titles motivate and enable women farmers to increase output by adopting improved techniques and practices. Equal property rights enhance women's bargaining power within the family and the community. Women owning land (even in small plots) are more respected by family members and villagers. They are able to challenge social inequalities and voice their concerns in public decision-making bodies. Land rights can improve gender relations (e.g., matrilineal inheritance). Hence, equality in land rights is linked with women's empowerment.

Bina Agarwal (1997) in the article "Field of her Own" asserts that the single most important factor affecting women's situation is the gender gap in command over property (arable land). Land is accessible mainly through inheritance. Inheritance laws vary according to religion and region. There is a gaping chasm between women's legal rights and their disinheritance in practice. This affects their welfare, efficiency, equality and empowerment.

Effective land rights remain elusive due to social, bureaucratic and ideological factors. Many patrilineal communities have dissonant (contemporary) inheritance laws and (traditional) marriage practices. There is a male resistance to endowing daughters with land. Women traditionally grant their share of the parental land to brothers thereby providing a form of traditional social security. Certain government functionaries obstruct the implementation of laws in women's favour. Prevailing biases, such as allotment of land to male household heads, affect the documentation of women's shares, court judgments etc. A systematic bias against women and female children is reflected in the intra-household allocation of resources. Where both men and women control resources (especially in poor households), women spend their income on the family's basic needs while men spend theirs on personal needs. The gender gap between legal rights and actual ownership is synchronous with ownership and effective managerial control. Female illiteracy, high fertility and female seclusion practices restrict women's activities, knowledge and skills. Additional barriers like taboos against women ploughing and gender bias in extension services also act against them. Current national statistics measure 'female headed households'. This category does not reflect actual women title holders. Krishnaraj reports that land ownership is recorded under the name of the household. Interviews of the wives of farmers (who committed suicide) in Vidarbha reveal that the plot of land is recorded in the name of the father-in-law (and not in the name of the deceased husband) (Datar, 2008).

There is a constraint on land resources in India because of population growth, agricultural expansion and industrial growth. This causes widespread environmental degradation and changes in land use patterns. Between 30–50 per cent of India's land area is wasted (Agarwal and Narain, 1989). Women in rural India are particularly affected because they are the primary collectors and users of natural resources. According to Agarwal and Narain, in India, most people subsist on a biomass-based economy. Almost half of the industrial output is derived from biomass-based industry. Hence, there is a need to increase biomass production with an increased attention to trees, grasses, standard crops, forests, grasslands and croplands.

Competing pressures on land resources have led to a wood fuel and fodder crisis. Sources of rural domestic energy include twigs, sticks, dung and agricultural residues (corn cobs, sugarcane husks, rice straw etc). A shortage of these residues contributes to the rural energy crisis. The rural poor (largely lacking private property) depend heavily on free collection of wood fuel and fodder from the commons. They also rely on common land for gathering food, raw materials for artisans and other subsistence needs (Jodha, 1986).

Of the 264 million hectares of land fit for vegetation, an estimated 84 million hectares comes under 'wastelands'. The term 'wasteland' implies land which has been degraded by either natural causes or poor soil and water management techniques (and which lies unutilized) (Menon, 1996). The Ministry of

Environment, Government of India set up the National Wastelands Development Board (NWDB) in 1985 to address growing environmental concerns. Its task is to formulate a social forestry policy for the regeneration of wastelands through local population involvement. State agencies and development organizations are involved in various activities like establishing nurseries, land preparation, afforestation and maintaining plantations. In return, the beneficiaries get usufruct rights to the plantation, employment (through nurseries and planting activities), fodder and fuelwood supply and a portion of the income from the sale of plantation products.

In the following paragraph a few illustrations (case studies) of sustainable natural resource use practices are presented.

SARTHI in Gujarat (Sarin and Khanna, 1993)

SARTHI (Social Action with Rural and Tribal Inhabitants of India) is based in Godhar village (in the Santrampur block of the Panchmahals district, Gujarat). It was started in 1980 though it became an independent entity only in 1985. One of SARTHI's major programmes in the area is the development of wastelands exclusively by women's groups. The programme is unique because it originated from the women of Muvasa village. They themselves articulated the need to eliminate fodder and firewood scarcity.

The Panchmahals district is one of the most industrially backward districts of Gujarat. About 89 per cent of the total workers in this district depend on agriculture and allied fields. While 53 per cent of land is under cultivation, only about 9 per cent is irrigated. The main source of irrigation is wells. Thus, over 90 per cent of the land under cultivation is dependent on rain. This leads to high occurrence crop failures due to deficient or untimely rains. Besides crop failures, acute fodder scarcity (resulting in heavy cattle mortality) causes large-scale seasonal migration to other areas. The district has a low literacy rate, high population density, less urbanization and a high percentage of tribals.

The forests in the area are prone to illicit felling, commercial exploitation, agricultural clearance and uncontrolled grazing. The status of women is low. In addition to domestic chores (cleaning, fetching water, gathering fuel, livestock maintenance, cooking, childcare, care of sick and elderly), many women work for daily wages. Desertion, domestic violence, alcoholism and husbands remarrying (to beget sons) are common problems. Women own little land or property. Decisions regarding land are almost exclusively made by the men. This makes it difficult to involve women in wasteland development. Women belonging to small and marginal farmer families struggle to meet fuel and fodder needs. Women have to traverse long distances to procure fodder and fuel (especially during summer).

In 1988, SARTHI initiated the development of private wastelands with a few families in Muvasa village. The goal was to increase fuel and fodder availability for women. The field staff explored the possibility of a group developing a piece of common land together. Women were interested in this endeavour as it would reduce scarcity issues while simultaneously creating employment (in the form of daily wages). The ground rules were mutually worked out. These included involvement of all members in land preparation, planting work, protection of the plantation, right to an equal share of the produce, collective decision making and only women being eligible for group membership. The last rule was established because women bore the brunt of fuel and fodder scarcity. A diverse range of species (selected by the women themselves) were planted. The plantation created many income and employment benefits. After its success, none of the members worked on owned land or public lands. The plantation supplied their daily and yearly requirement of fodder and firewood. Their work timings being flexible, they could allocate more time to completing household chores. They earned more on the days they put in more labour. Due to the flexible working hours, women could also go to dairies to sell milk.

Many problems were faced during the initial phase of work. Men of different castes and classes questioned the motives of the organization. They felt that housework was being neglected and hence objected to women attending meetings. When planting was started, few village men pressurized the women to plant only eucalyptus trees. Women refused to succumb to these demands when they were reminded of their primary needs for fodder and fuel. SARTHI staff dealt with men's hostility by calling a meeting of the entire village and explaining the purpose of working with a women's group. This allayed men's fears. Apart from distributing cattle and deepening wells, SARTHI conducted environmental awareness camps and tried to involve men in other activities of the organization.

Women's involvement in wasteland development is related to their resourcelessness in terms of land ownership. Their secondary social status denies them opportunities to act equally in non-domestic domains. Under SARTHI, women's empowerment through group involvement indirectly strengthens them to fight against familial injustices. The wasteland groups created a new social space for women.

Deccan Development Society (Gayatri Menon, 1996 and Anandi Team for Agakhan Foundation)

The Deccan Development Society (DDS) is an NGO which has been working with poor, scheduled caste women's collectives in over 60 villages of Medak district (Andhra Pradesh) since 1983. It recognizes that access to land and land-based resources is the basis of survival for poor communities. They also foreground the primacy of land in determining class and gender relations. DDS enables poor women (from landless and marginal landholding families) to establish their claims on land resources. This is achieved through a variety of government schemes, partnerships and linkages (including land purchase, land lease, community grain fund and wasteland development).

Its strategy is based on the premise that processes which enable poor women to access land have to be collective and multi-branched. These processes must explore new institutional arrangements which give land rights to women. Individual women cannot tackle the patriarchal assumptions inherent in social customs (which govern the inheritance and ownership of land). They cannot resist state policies on land and land based resources individually. The *'sangam'* or collective endeavours to give women access to resources. Women are empowered through the formation of sangams or women's cooperatives.

When DDS began working with women, it gradually understood the level of poverty that marginalised women were grappling with. DDS started women's groups (called sangams) and initiated thrift activities. Women's sangams operate differently. Women are more honest and request smaller loans for basic household needs. They rarely default on their loans. These forums enable women to share their experiences, problems and insights. They foster a sense of collective identity and trust.

A community-level space was created by positioning sustainable livelihoods as the primary objective of the strategy. This is based on the recognition of women's roles in mobilizing and managing community resources. Changing women's situation (of subordination) was linked to improved life for the entire community. Therefore, it was accepted by the men and the community as a whole.

In the areas where DDS operates, the community consists of socially and economically marginalized castes. In this area, even the poor own a small piece of land. This is a direct consequence of the feudal system where workers were given land as gifts or inams by landlords. Furthermore government sponsored land reforms have dispersed land into the hands of more people. Since landholdings are meagre and often of poor quality, most dalits work as wage labourers on farmland or in sugar factories.

Activities of DDS in natural resource management and its sustainable development initiatives include:

- Watershed management programmes.
- Health-related programmes such as 'The Medicinal Plants Project'.
- Balwadis for the children of working women.

- Education through the Green School or 'Pachasaale', nights schools, Krishi Vigyan Kendras (KVK).
- Food security for communities through agriculture and allied practices. DDS has taken four steps to secure food security:
    1. Enhancing the productivity of marginal lands owned by dalit women thereby increasing the food available to their families.
    2. Encouraging sangam women to lease lands from bigger farmers, cultivate them collectively and share the produce. This programme combines the principles of empowerment and enhanced food security.
    3. Pilot an Alternative Public Distribution System (Community Grain Fund) in 30 villages on the principle of local production, storage and distribution systems. This programme established a locally managed, community-controlled public distribution system.
    4. Initiating a Community Gene Fund through which women started an in situ conservation of local landraces. They also developed a village level germplasm bank of traditional crop varieties. Through this programme, women established control over their own seed systems.
- Collaboration with government officials and agencies such as the Indian Council of Agricultural Research to set up Krishi Vigyan Kendras (Agricultural Science Centre) which help in the implementation of government-sponsored watershed development projects.
- Women's media, e.g., Sangam Radio.
- Other activities like campaigns, workshops, biodiversity festivals, eco-enterprises (e.g., Vermi Compost programme), research studies, organic markets etc.

Water

Water is a key resource in ecosystem management. The importance of water in India is rooted in two reasons both of which are related to the monsoon pattern. Short concentrated wet spells and a dry, long scorching summer desiccates and volatilizes organic matter. It turns soil into fine dust thus eliminating the continuity of biomass cover and mulch. The monsoon pattern also renders the soil vulnerable to erosion.

In this situation, large centralized water systems have come to govern water management practices in India. Unfortunately, they have deepened the pre-existing disparities of Indian society. Such practices ignore the two important principles of judicious water distribution. Limited water norms and equitable access to water should be synchronized to ensure livelihood needs. These needs should be treated equivalent to the employment guarantee and the right to work. According to Datey, regenerative practices require a small but definite quantity of water at the right time of crop growth.

Judicious water allocation (for different needs) depends upon the orientation of the policy makers (patriarchal, capitalist/market-oriented or 'ecologically'-oriented). The recent water policy of the Government of Maharashtra cites drinking water as first priority. Industrial needs occupy the second rung while agriculture is relegated to the third slot. Datey asserts that a participative approach is necessary to evolve common understanding. However, this approach has to be based on a scientific appraisal of resource availability and constraints. These constraints should be overcome by an optimal and cooperative utilization of resources.

If this approach is not assimilated in the commonsense understanding of water-deprived people and they do not pressurize the government (to integrate these principles in its policies), it will remain at an experimental level. Guha points out that this is a matter of political ecology, i.e., the politics of distribution of natural resources. A few models of sustainable use of water in India are discussed below.

The Pani Panchayat Model of Water Management for Irrigation (Salunkhe, Lokur and Pangare, 1997)

Land and water are equally important for agriculture. However water ultimately determines the economic status of a farmer. Land ownership stratifies rural society into rich and poor classes. Generally, in irrigation projects, water is allocated in proportion to landholdings – farmers with large landholdings get more water. Furthermore, farmers whose fields are closer to the water source get greater benefits. This increases the gap between big-rich and small-poor farmers. Many irrigation projects (like dams, percolation tanks, canals etc.) have failed to achieve their objectives. This is linked to their failure in redistributing resources among the rural population. Ramchandra Guha states that while political economy studies economic distribution conflicts, political ecology studies ecological distribution conflicts. It appears that these conflicts run concurrently (Guha and Martinez-Alier, 1998: 22).

To overcome these social injustices, the Gram Gourav Pratishthan (GGP), which establishes lift irrigation schemes ensures that the planning and implementation of irrigation schemes is just. The term 'Pani Panchayat' was coined to denote a five-member committee. This committee monitored the first lift irrigation project established by the GGP. The term 'Pani Panchayat' refers to the mobilization of farmers' groups to formulate and implement community irrigation projects. Today, it symbolizes the principles of equitable water distribution.

Lift irrigation schemes enable lifting water through the canal system. This water is transferred to upland areas thereby increasing the efficiency of water utilization. Using this method, people who have lost land due to the construction of dams can access water. Thus, it enables extensive irrigation. A limited amount of water is distributed on a 'per person' basis leading to judicious water utilization. The most efficient way to utilize rationed water is to grow food grains. This norm deters the cultivation of water-intensive cash crops (like sugarcane).

Migration can be prevented by providing employment at the village level. This is best achieved by making water available for irrigation. In Pani Panchayat villages, farmers have returned from cities to practise agriculture (after the lift irrigation scheme was adopted). People displaced by the construction of dams frequently settle down at the edge of the water. However, they cannot use this water for irrigation as their lands are at a higher level. Pani Panchayats solve this problem (by lifting water) and enable the rehabilitation of people in familiar surroundings. The Pani Panchayat Lift Irrigation Scheme also provides water for drinking and domestic use.

The Pani Panchayat follows certain pre-determined principles. The irrigation schemes are implemented for groups of farmers rather than individuals. They believe that water is a common property resource and that all villagers (irrespective of land ownership) should have equal access to it. Membership is offered to landless villagers to enable them to become share-croppers to farmers having more land. Therefore, this scheme promotes gainful employment in the village itself. Water is allocated on the basis of family size (a maximum of five members per family) rather than on the size of landholding (at the rate of half an acre per family member). The maximum share per family unit of five is two and a half acres. To prevent absentee landowners from manipulating land prices, water rights are detached from land ownership. The Pani Panchayat schemes are located in drought-prone areas. Water is used for cultivating seasonal crops to discourage the cultivation of crops which require perennial irrigation and large quantities of water (specifically sugarcane).

To inculcate responsibility towards the project among members, the participants are expected to contribute 20 per cent of the total cost. For examples, the schemes in Purandhar and neighbouring talukas irrigate 3,000 acres of land. Thus, they support a population of 10,000 people from 1,500 families. The total capital investment was 75 lakh out of which people contributed 15 lakh.

The first Pani Panchayat Lift Irrigation Scheme was planned in 1980. It was based on a plan known as *panch-teen-do* or five-three-two. This represents the ratio in which the money would be raised (from three sources) – a 50 per cent subsidy from the government, 30 per cent interest-free loan from the GGP and 20 per cent by farmers in cash and voluntary labour. In 1981, GGP revised the cost sharing ratio due to government policy and redtapism. The 50 per cent subsidy of the government was eliminated while the contribution of the beneficiaries was increased to 30 per cent in cash and 10 per cent *shramdaan* or voluntary labour. The remaining 70 per cent was provided by the GGP as an interest free loan to be paid in three years.

Certain factors facilitate the implementation of lift irrigation schemes. GGP schemes are group endeavours which ensure equitable distribution of created water resources. This perception of equal benefits motivates farmers to form groups and implement the Pani Panchayat scheme. GGP assists only those farmers who approach the organization as a group. The first contact is initiated by the group and not by GGP. The group also contributes the majority of the project costs including the cost of the initial technical survey. This ensures a sense of responsibility towards the project. Voluntary labour or *shramdaan* ensures that the farmers feel involved (in their own welfare) rather than being the recipient beneficiaries of a development scheme. Importance is given to local leadership to keep the group together. Leadership is developed to emphasize group initiation and management. These schemes are informal groups managed by the members themselves. They are not registered as societies, trusts etc. Thus, the members are not answerable to any authority outside the village. The avoidance of elections prevents groupism and conflict. This also eliminates free riders. The GGP provides necessary technical inputs and guidance. It also trains the villagers about the maintenance of the system.

The resultant stability in agriculture has increased the income of families within the Pani Panchayat Lift Irrigation Scheme. The positive impact of fund availability is discernible in the lives of women. It has enabled women to visit village fairs, weddings and other social functions. This has increased social interaction and created greater awareness. In the Madhapur Lift Irrigation Scheme, women asked for training to increase their knowledge and skills in agriculture. Although women are not involved in the day-to-day functioning of the schemes, they have contributed to starting them (through their economic contribution). Women raised money by selling jewellery, working longer hours or renting their plots. Economic stability and increased awareness have made them more vocal and confident. They have begun to express their needs and are willing participants in community level decision-making. In many cases, women supported their husband's decision to participate in the irrigation scheme. The GGP plans to take up specific activities for women (like training) especially since the request has come from women themselves. Assured irrigation has brought greater food security in the combined form of food grains and cash crops.

Sustainable water development and management should be an integral part of rural development in India. Unfortunately, the political leadership in Maharashtra refused to adhere to 'limited water use'. Hence, they did not adopt a policy based on this principle. It is essential to manage existing irrigation potential in a rational manner. Concerted efforts are needed to stabilise rain dependent agriculture. Rural communities can effectively handle resources if they are allowed to manage them locally. It is easier to introduce women's interests in innovative management systems because the groups involved are small. The principles involved in management are specific to the system. This accommodates the special interests of different resource users in the community.

Uthhan Model of Drinking Water Management (Barot, 1997)

Another example of resource-poor communities (especially women) conserving and managing local natural resources is illustrated in Gujarat. Within this state, the scarcity of drinking water is widespread, chronic and deep rooted (Barot, 1997).

The Gujarat Water Supply and Sewerage Board (GWSSB) was established to instal and run regional schemes. The Seventh Five-Year Plan (1985–90) introduced desalination and de-fluoridization plants. The Eighth Five-Year Plan (1990–95) acknowledged the need to recharge groundwater. It advocated rainwater harvesting and conservation through the Narmada dam. People's participation was highlighted for the first time. This process clearly reveals the centralization of water systems in terms of technical management and solutions. It depends on underground water and on irrigation dams to meet drinking water needs. This was unsuitable due to various reasons. The pipelines supplying water from the dam stretched over long distances. Head reach villages had more control over water and diverted it for other uses. It increased dependence on electricity (for pumping) and on carelessly run diesel sub-pumping stations. Due to the dependence on bore wells, the water output decreased and quality deteriorated. The water distribution systems were located near houses of prosperous people who denied access to the disadvantaged. The pipeline was poorly maintained and the communities were not involved.

Utthan facilitated the emergence of a community-based group called Mahiti during 1981–1994. These organizations started resource planning and management in the Bhal region of Gujarat. This region covers parts of Ahmedabad, Kheda, Surendranagar and Bhavnagar. The geo-climatic environmental conditions of this region are very harsh. The groundwater is shallow and saline. This region is near the sea and experiences erratic monsoons. The source of the drinking water pipeline was about 100 km away from tail-end villages. The scheme was initially meant for 52 villages. However, it was later stretched to include more than 75 villages. People fought violently while collecting water from the 'sump' (underground water collection tank). Women were major victims because they (and their children) were sent to procure water to avoid fights. Water was supplied through pipelines or tankers with 4–5 litres allotted to each person each day. The community's requirements for cattle, washing, bathing, etc., were ignored. Communities resorted to digging 'virdas' (pits) or dry shallow ponds. People had no choice but to guard these pits or walk 10–12 km to get water during summer.

Women collected water from puddles (during the monsoon) and from village ponds (in winter). The ponds would become saline due to the seepage of sea water and a highly saline ground water table. Initially, the village ponds were maintained by village panchayats and the villagers themselves. But, the advent of pipelines led to their disempowerment. Powerful villagers in the head reach took away more water. The poor quality of installed pipelines caused frequent breakages. These breakages polluted treated water. Scarce electricity caused water to flow with very low pressure of water. The quality and quantity of water deteriorated in summer. Tail-end villages had no control over the quantity or quality of water.

The Utthan and Mahiti teams evolved alternatives to make saline land productive. They established income-generating activities apart from making drinking water available. This led to a woman's and (later) a community's movement. These movements empowered them to solve their drinking water issue. The women in Bhal pressurized the Gujarat Water Supply and Sewerage Board to approve decentralized rainwater harvesting structures (like plastic lined ponds, roof water collection tanks etc.). The water movement has now spread to other parts of Gujarat and is changing government policy.

The intervention of Utthan and Mahiti had a positive impact on the region. The increased quantity of water reduced the drudgery of women. Health improved due to hygienic conditions, reduced diarrhoea cases, increased attendance of children and teachers in schools etc. There was a sense of ownership and self-reliance towards the resource. Many women evolved a 'holistic' water plan for their own villages (e.g., roof water tanks for human consumption, community pond systems for bathing, washing etc.). Initially women were restricted by social pressures from within their communities. They faced ego-related problems of sarpanchas, village leaders and their own men. In 1987, many issues were taken up by women's groups in Bhal region. Traditional moneylenders' houses were raided by anti-corruption squads. As a result, many families paid back the original amount lent to them and became debt-free. Women procured salvadora (piloo) trees on lease from the revenue department. This protected trees from direct grazing and (by direct marketing) removed exploitative middle men. They also acquired saline wasteland under women's collectives. Many organizations identified with the women's water movement in Bhal.

The sustainable livelihood paradigm assimilates local wisdom and knowledge. Scientific inputs can improve traditional systems and sources. Livelihood priorities can be managed on a sustainable basis if their overall management is decentralized. Decentralized drinking water systems offer a greater potential for improvement, development and the equitable distribution of resources. Local communities can be involved at all stages of its management and utilisation. For this to happen it is be essential to change the institutional structures and the roles of authorities which control such systems. Almost all decentralized drinking water systems which take the gender perspective into consideration are found to be economically sustainable. Women's involvement brings out the needs of all groups in the community. Women employ innovativeness based on traditional wisdom, knowledge and value systems.

Hivre Bazar: A Model Watershed Experiment (Menon et al., 2007)

Hivre Bazar is a village in Nagar taluka in the Ahmednagar district (Maharashtra). It was chosen for the 'Adarsh Gaon Yojana' (AGY) in the early 1990s. It was based on the model of Ralegaon Siddhi (under the leadership of Anna Hazare). This model was an example of participatory and sustainable watershed management. The model inspired the formation of the AGY. Hivre Bazar is unique because the initiative is partly rooted in state support for a community-based natural resource management (CBNRM) approach. It exemplifies a village community at the forefront of decentralized natural resource management without the 'outside' intervention of an NGO.

Background of Hivre Bazar

It is bounded by hillocks on three sides and hence is ideal for watershed development. The main village settlement is at the foot of the hills and towards the upper reaches of the watershed. The village is divided into three sub-watersheds. The agricultural land is on the upper reaches of the watershed. The availability of water in the lower reaches is higher because of a check dam constructed in 1982.

Around 81 per cent of the land is cultivable. Out of the remaining non-cultivable land, 7 per cent is forested and 4.5 per cent is a wasteland. Jowar (sorghum) and Bajra (pearl millet) are the main crops. The village is homogeneous in its social composition with all castes (SC and ST) engaged in agriculture. The Maratha-Kunbi caste is the dominant community. They own approximately one hectare land for every 20 hectares of available land. The dalits do not own land though some of them are sharecroppers.

Due to water scarcity agricultural productivity was often poor (with only a single crop). Landless, marginal and small farmers were the worst affected. Migration for work was the preferred option. Medium farmers subsisted on the Employment Guarantee Scheme and by working as wage labourers outside the village. Drinking water was scarce during summer and people had to wait for government tankers. The rainwater was lost due to the absence of vegetative cover and structures to harvest water. Thus, even in a year of normal rainfall, wells in the village dried up by December or January. Fodder and firewood were in short supply. There were no attempts to regulate open grazing and cutting of trees.

The success of AGY and CBNRM in Hivre Bazar is attributed to the dynamic leadership of their sarpanch Popatrao Pawar who was elected to the gram panchayat in 1989. He addressed the issue of drinking water by convincing the Panchayat Samiti to instal new handpumps. He proceeded by renovating the village school and appointing honest teachers. The village temple was also renovated.

In the early 1990s, he was influenced by Ralegaon Siddhi and Anna Hazare. In 1993, the Social Forestry Department (in Ahmednagar district) created 40,000 contour trenches to preserve 70 hectares of forest land. Various indigenous trees (like babool, tamarind, shisham, khair) and fast growing exotic varieties (like gliricida, eucalyptus, Australian babul) were planted. Grazing was banned in the area. However people were allowed to graze their animals on adjacent private wastelands. People had to contribute Rs 100 to the village fund for securing fodder from the forest. The trenches had a delayed but visible effect. By the end of the monsoon, many adjacent wells were still full and the area they served increased.

Work under the AGY began in 1994–95. The Gram Sabha started the project by forming an NGO (Yashwant Krishi Gram and Panlot Vikas Sanstha) comprising local villagers. Sub-committees were formed in each watershed. The NGO employed technical staff to plan and prepare technical reports. They recruited volunteers for organizational work. The plans were discussed in the gram sabha and opinions were sought from people.

All soil and water conservation works in Hivre Bazar were implemented on common lands or private wastelands/grasslands. Continuous contour trenches (CCTs) were dug on hill slopes to arrest soil erosion, harvest water and encourage the growth of grass. A number of water harvesting structures (like check dams, percolation tanks, loose boulder structures etc.) were built in the village. The programme included plantations on forest lands and roadsides.

Some other bans and measures in the village (which were not part of the AGY) were added later. These were central to watershed development in Hivre Bazar:

- Ban on bore wells for irrigation, cultivating sugarcane and banana and selling one's land to outsiders.
- Social measures – anyone marrying in the village needed to undergo the ELISA or AIDS test.
- The four main bandis or bans were:
    - 'kurhad' bandi – ban on felling trees
    - 'charai' bandi – ban on tree grazing
    - 'nasbandi' – family planning
    - 'nashabandi' – ban on liquor (was supported by many especially women)
- The main aim of watershed development in Hivre Bazar was to increase water availability (for drinking and irrigation) and to improve agricultural productivity.
- Various measures were adopted to impound water. They helped recharge ground water. This was noticeable in the

water level of wells and in the two-fold increase in number of wells in the last 10 years. In a year of normal rainfall there is enough water to irrigate kharif bajra and rabi jowar along with some summer vegetable crops. People currently grow two to three crops in a year of normal rainfall.

- Increased soil moisture in non-irrigated land helped to increase productivity. This allowed the cultivation of diverse crops like potatoes, onions, grapes, pomegranates, flowers and wheat.
- Wage employment opportunities have increased. The demand for labour has increased due to increased intensity of agriculture and multiple seasons. People migrate lesser than in the past as many households now grow a second crop.
- Other activities like land improvement by individual farmers, construction of buildings, setting up pipelines etc. have emerged. Through increased shramdaan, villages can save a significant component of the subsidy thereby building up reserves.
- Restrictions on grazing were rotational in the initial stage. Some private lands were kept accessible as grazing grounds.
- Livestock rearing: Stall feeding and cows of better breeds (jersey) have been introduced. Stall feeding is encouraged to protect newly planted trees. This helps in the natural regeneration of vegetation and prevents soil erosion due to open grazing. It also increases the production of organic manure. Animal husbandry has become an important livelihood option.
- Self-Help Groups (SHGs): These were introduced to generate savings and improve overall well-being of women. They have met with limited success as agriculture is more prosperous.
- Wage employment opportunities have increased for the landless poor. This is partly due to the increased prosperity of landed individual which increases labour requirements. Many landless and small farmers have invested in cattle.

The Narmada Bachao Andolan (NBA) (John R.Wood, 2007)

NBA questions the establishment of big dams (which have environmental and human repercussions). The conflict between the Narmada Bachao Andolan ('Save Narmada Movement', NBA) and the Government of Gujarat started with the improper rehabilitation of people dislocated by the Sardar Sarovar dam. This dam is being built on the border of Gujarat and Maharashtra. The World Bank had initially promised a big loan for this project. Later, an independent review commissioned by the World Bank expressed doubts over the project. The review recommended that the Bank reconsider its funding support.

NBA's campaign agenda included the protection of the environment of the Narmada valley. It fought for the human rights of the 'oustees' (i.e., people being displaced by the dam's huge reservoir and canal system). The treatment meted out to these oustees (mostly adivasis) became their special concern and the focus of their attack. The NBA linked the environmentalists' struggle (to preserve forests) to the preservation of the adivasi way of life. The NBA also questioned the development discourse as a whole and the policy decisions behind big dams in particular.

Profile of Sardar Sarovar Dam

In 1969, the Government of India (under Mrs. Indira Gandhi) constituted the Narmada Water Disputes Tribunal (NWDT) to resolve the inter-state water dispute between Gujarat and Madhya Pradesh. While Madhya Pradesh proposed a dam height of 210 feet, Gujarat demanded a height of 530 feet. In 1979, the Tribunal compromised between the claims of the two states. Accordingly, they allotted 9 MAF of water to Gujarat and a dam height of 455 feet (138.7 m). This would create the world's longest cement-lined canal (a total stretch of 532 km through central and northern Gujarat and into Rajasthan). It was projected that 1.8 million ha of land in Gujarat and 75,000 ha in Rajasthan would receive water from the Sardar Sarovar Project. Drinking water would become available for 3,800 villages and 147 towns in Gujarat. Full-scale construction of the Sardar Sarovar began in 1987.

Human Rights Issues

The people affected by the construction of the dam are rarely asked to voice their opinion on the project, its location or its effects. The plight of displaced people has led to questions of evacuation, inadequate financial compensation, resettlement and rehabilitation. Whole villages and kin groups will have to be moved together to prevent social breakdown. They will have to be given certain facilities to enable them to start new lives. The impact on 'host' communities of large numbers of displaced people in the resettlement area is indeterminate. Aboriginals (or indigenous or tribal people) form a significant proportion of the displaced population. Their lives depend on the surrounding forest habitat. The landed will profit from the year-round water supply while those who are landless will not.

Environmental Issues

These issues are related to interrupting the normal flow of a river. This affects the environment downstream. Stoppage or severe reduction of river flow adversely affects vegetation, fish and other estuarine life. The upstream migration of fish (for spawning) is affected. The sedimentation pattern of the river is altered. Silt accumulates in the reservoir till the dam's storage capacity is exhausted. This reduces the life expectancy of the dam and creates the danger of flooding during storms in the rainy season. The problems caused by massive water storage in the dam's reservoir are a major area of concern. Valuable land and forests along with associated flora and fauna get submerged. This adds to deforestation and its effects. Massive water storage is accompanied by a marked increase in water-borne diseases, especially malaria, schistosomiasis, encephalitis etc. Damage due to earthquakes and the likelihood of flooding downstream are other concerns.

Cost-benefit Analysis Issues

NBA states that the Sardar Sarovar Project was defying the cost-benefit norms of the development perspective. A standard cost-benefit analysis measures (in a specified time-frame) the cost of dam and canal construction, submergence, rehabilitation, catchment area treatment, command area development, drainage, repairs and maintenance of project facilities. The sum of these is weighed against the benefits for agriculture (through irrigation) and the use of water for non-agricultural purposes (like hydro-electric power). The net benefit is calculated by subtracting the value of command area productivity in a 'with-out-project' scenario from that of a 'with-project' scenario. Indirect benefits such as employment generated by the project, employee spending and production in various sub-economies are also calculated. Other costs under environmental costs (like loss of trees, vegetation, soil, flora and fauna) and their effects on humans and environment have to be calculated. Human costs in the form of losses and gains for displaced people (depending on their response to rehabilitation), its effects on their children and their future generations also need to be factored.

The Narmada Bachao Andolan is accused of 'leading nation towards barbarism' and 'belonging to anti-technology school' which does not believe in productivity enhancement and labour saving. However, one must admit that in the vision of decentralized communities, provided by Medha Patkar, women find more space to highlight their concerns for reproduction, subsistence and regeneration.

13

Conclusion

Ecofeminism provides a paradigm shift from other strands of feminist theories like liberal, socialist, developmentalist and postmodernist. It refutes the dualist understanding of nature and culture, i.e., nature associated with women and culture with men. It attributes this dualism to patriarchy which used this dualism conveniently to generate "scientific methodology and knowledges" to relegate women to invisibility. In the process, it reduced both nature and women to an "inert and static status".

There are different strands in ecofeminism from liberal to socialist and subsistence perspective, but it is the cultural ecofeminism taken as a representative of them all and reduced to accusation of 'biologism' and 'essentialism' by 'mainstream feminists'. In this book I have highlighted the ecofeminism based on materialist conception of history of man-woman-nature relations which rejects capitalist patriarchal theories of development that promote 'mainstreaming of gender' thereby threatening the survival of life on earth.

Feminism defines liberation in terms of developing capacity for self-determination, scienticism and instrumentalism, i.e., liberation of mind from patriarchy. Ecofeminism questions development process which continues to exploit, oppress, and resort to violence and colonization. It argues for decentralization, for self-sufficiency, autonomy, mutuality, self-provisioning, subsistence relations instead of profit relations and to maintain biological and cultural diversity. Ecofeminism wants women's activities of caring, producing food and concern for food security to be brought centre stage in the alternative models of development process. It encourages women to be aware of their regenerative capacities (not just mothering and child bearing) and identify with the regenerative processes of nature so that they would fight for the sustainability of ecological processes assuring survival of humankind dependent on natural resources. It rejects the sole concern of growth within the destructive paradigm of capitalist patriarchy. It also believes that processes of individuation and modernism should not be accomplished at the cost of mutuality and respect for traditional knowledges, especially of women.

In the process of exploring alternative models of development I have highlighted that simple solution of going organic and substituting non-renewable energy sources such as wind and solar energy is not sufficient. To do away with extreme disparity in resource use and concentration of power to control natural resources, we need to develop more dispersed sources of energy and have to address concern of unequal energy exchanges between developed and developing countries and the rich and poor within the same country. Biomass-based strategy has been highlighted in the chapter on alternatives. Subsistence perspective advocated by Maria Mies and others sounds unrealistic but points out sharply the need to address issues such as how to achieve decentralized production systems. Ecofeminism cannot alone change the world, but it would focus on women's needs and women's agency to contribute to the ecologically sustainable world.

Notes

1. **The book *Silent Spring***

   The publication of *Silent Spring* in 1962 by Rachel Carson is credited with initiating the environmental revolution. It listed the adverse effects of the misuse of synthetic chemical pesticides and insecticides. *Silent Spring* generated a lot of controversy and heightened public awareness. Carson warned that the use of chemicals to control insects was interfering with the natural defences of the environment. Carson first became concerned about the effects of DDT in 1945 while working for the US Fish and Wildlife Service.

   Silent Spring took the pesticide issue out of academic circles and technical journals. It was an ecological book which brought this issue into the public arena. *Silent Spring* propelled people into action against the misuse of chemical pesticides. It instigated changes in local and national government policy in the US and several European countries (UK, Sweden, Denmark, Hungary).

   [From McCormick John (1989), *The Global Environmental Movement: Reclaiming Paradise,* Belhaven Press, London, UK]

2. **Three Mile Island**

   A nuclear power plant accident which occurred on March 29, 1979 at the Three Mile Island plant in Pennsylvania.

3. **Externalities**

   Beneficial or harmful effects associated with the production and consumption of a good which are not included in its market price.

4. **Jayaprakash Narayan, Veteran Gandhian and Sarvodaya Leader**

   Jayaprakash Narayan did not join formal politics after independence. He continued doing social work and maintained his moral authority as an independent thinker. As a response to corruption and continuing poverty during the early 1980s, he announced a movement called Sampurna Kranti (Total Revolution). This movement was joined by many young men and women who continued to agitate for two-three years. It ended with the declaration of the emergency (announced by the government led by Indira Gandhi on 26th June, 1975) which lasted till 1977. Many oppositional leaders were arrested. Jayaprakash inspired the youth who organized themselves under different banners (like the Chhatra Yuva Sangharsh Vahini in Maharashtra and Bihar; Navanirman Sangharsh Samiti in Gujarat etc.). Some of them remained active in rural development and environmental activities. They also rallied around the Narmada Bachao Andolan.

5. **Gandhi and Nehru**

Many environmentalists believe that the Gandhian vision was cast aside by Nehru who imposed on independent India his model of capital and energy intensive, environmentally-destructive economic development. While Gandhi's vision of free India was centred on village renewal, Nehru's vision focused on rapid industrial development. Gandhi preferred stability to change but Nehru preferred change to stability. The former's value system was based on religion; the latter's on science and scientific temper. Nehru stated that sufficient food, clothing, housing, education, sanitation etc. for every Indian was the chief goal of economic planning. Both Nehru and Gandhi agreed that the elimination of poverty was essential goal. Nehru was convinced that this could be achieved only through rapid industrialization and the use of modern technology. Thus, the distinguishing characteristic of the Nehruvian model is an alliance between science and technology with the development process.

[From Ram Chandra Guha and J. Martinez-Alier (1988), *Varieties of Environmentalism: Essays North and South,* Oxford University Press, New Delhi]

6. **Earth Summit**

The United Nations' Conference on Environment and Development which took place at Rio de Janeiro (Brazil) in June 1992.

7. **Brundtland Commission**

The World Commission on Environment and Development chaired by the Norwegian Prime Minister Gro Harlem Brundtland. The Commission's report *Our Common Future* (1987) popularized the notion of sustainable development.

8. **Green Revolution in India**

In 1970, Norman Borlaug was awarded the Nobel Peace Prize for developing high-yielding varieties (HYVs) of wheat (also called "miracle seeds"). "Green Revolution" is the science-based transformation of Third World agriculture. Punjab in India was its most celebrated success.

In the 20th century, from World War I to independence, Indian agriculture suffered due to various factors like land alienation, forest reservation, expansion of cash crop cultivation (introduced during colonialism), reduced exports due to worldwide recessions and depression coupled with the chaos of partition. Till the 1960s, India successfully pursued an agricultural development policy. This policy was based on strengthening the ecological base of agriculture and the self-reliance of peasants.

Land reforms were initiated in most states to fix reasonable rents, secure tenure for tenant cultivators and to abolish the zamindari (landlord) system.

Alarmed by growing peasant unrest in the newly independent countries of Asia, organizations like the World Bank, the Rockefeller and Ford Foundations and the US Agency for International Development attempted to stabilize these regions by intensifying agriculture. In 1961, the Ford Foundation launched an Intensive Agricultural Development Programme in India through the introduction of modern intensive chemical farming. The Green Revolution was officially launched in India in 1967–68. Dr M.S. Swaminathan is cited as the founder of Green Revolution in India. This mode was characterized by the use of HYVs or "miracle" seeds, intensive irrigation, large-scale use of pesticides, fertilizers and monocultures of crops. This led to reduced genetic diversity, increased vulnerability to pests, soil erosion, water shortages, reduced soil fertility, micronutrient deficiencies, soil contamination, reduced availability of nutritious food crops for the local population, displacement of large numbers of small farmers from their land, rural impoverishment and increased tensions and conflicts.

Intensive irrigation created a need for large-scale storage systems. Thus, they centralized control over water supplies and led to local and interstate water conflicts. In Punjab, farmers are actively campaigning to halt the construction of the Sutlej-Yamuna Link Canal. This canal will redirect water to Haryana to irrigate 3,00,000 hectares of land using Green Revolution techniques. However, local politicians in Haryana are lobbying for its completion. The worsening lot of the peasantry in Punjab (mainly Sikhs) has contributed to the development of Punjabi nationalism.

9. **Ecological Economics: Distribution, the Discount Rate and Incommensurability**

From an ecological point of view, the economy lacks a common standard of measurement because value depends upon the endowment of property rights, distribution of income, strengths of environmental movements and the distribution of power. These factors influence the allocation of prices to environmental resources and services. Such allocation is complicated by the rate of discount needed to weigh future costs and benefits.

Distributional issues invade the discount rate through the economic measurement of sustainability. The appropriate discount rate should be determined by 'sustainable' productivity of capital.

Incommensurability implies the lack of a common unit of measurement. These values cannot be compared on different scales. Incommensurability is a part of the tradition of ecological economics since Otto Neurath and William Kapp.

[From Ram Chandra Guha and J. Martinez-Alier (1988), *Varieties of Environmentalism: Essays North and South,* Oxford University Press, New Delhi]

10. **The Frankfurt School**

The term 'Frankfurt School' describes both a group of intellectuals and the 'Critical Theory of Society'. This theory was derived in the formative (and most radical) years of the Frankfurt School in the 1930s. Max Horkheimer, Herbert Marcuse, Erich Fromm and Theodore W. Adorno contributed to this theory. Their work played a significant role in the student (anti-authoritarian) movement of the 1960s.

[From Phil Slater (1980) *Origin and Significance of the Frankfort School: A Marxist Perspective*]

11. **Debiprasad Chattopadhaya**

Further Reading: *In Defence of Materialism in Ancient India* (1989), People's Publishing House, New Delhi.

12. **D.D. Kosambi**

Further Reading: *Combined Methods in Indology and Other Writings* (2002), compiled, edited and introduced by Brajadulal Chattopadhyaya, Oxford University Press, New Delhi.

13. **Social Constructivism (Wikipedia)**

Social constructivism is based on specific assumptions about reality, knowledge and learning. It is important to understand the premises that underlie the models of instruction of social constructivism.

Reality: Social constructivists believe that reality is constructed through human activity. Members of a society invent the properties of the world (Kukla, 2000). For social constructivists, reality cannot be discovered. It does not exist prior to its social invention.

Knowledge: Knowledge is also a human product. It is constructed socially and culturally (Ernest, 1999; Gredler, 1997; Prat & Floden, 1994). Individuals create meaning through their interactions with each other and with the environment they live in.

Learning: Social constructivists view learning as a social process. Neither does it take place only within an individual nor is it a merely passive development of behaviours shaped by external forces (McMahon, 1997). Meaningful learning occurs when individuals are engaged in social activities.

14. **Hermaphrodite**

An individual in which reproductive organs of both sexes are present; an organism, as an earthworm or plant, having normally both the male and female organs of generation; a person or thing in which two opposite qualities are combined.

15. **The Tenets of Deep Ecology**

The term 'deep ecology' was coined by the Norwegian philosopher Arne Naess. Here the author refers to the American variant of deep ecology. Followers of deep ecology accept certain fundamental assumptions about human-nature interactions:

Deep ecology argues that the environmental movement should shift from an anthropocentric to a biocentric perspective.

It focuses on the preservation of unspoilt wilderness and the restoration of degraded areas.

There is a widespread invocation of Eastern spiritual traditions as the forerunners of deep ecology.

Deep ecologists, despite their internal differences, believe that they are on the 'leading edge' of the environmental movement.

[From Ram Chandra Guha and J. Martinez-Alier (1988), *Varieties of Environmentalism: Essays North and South,* Oxford University Press, New Delhi]

16. **Energy and Power**

Energy is measured in different units in different systems of scientific measurement: calories, joules, BTUs (British Thermal Units) etc. However, kilojoules (KJ=1000 joules) and kilocalories (kcal) are the most frequently used units. For example, the human heart consumes about 1 joule of energy per heartbeat. The amount of energy required for making a cup of tea is about 75 KJ. If 1 kg of wood is burnt, about 16,000 KJ of energy is released. Burning 1 litre petrol releases about 32,000 KJ of energy. The basic unit to measure power (i.e., the rate of expending energy) is watt. 1 watt is equivalent to spending 1 joule energy every second. More prevalent units are kilowatts (kw) and hp (i.e., horse power = 746 watts). In urban life, energy is used predominantly in the form of electricity. Electricity consumption is measured in kilowatt-hour. Usually, kwh is called 'unit' (1 unit = 1 kwh = 3,600 KJ).

[From Wagle Subodh (1998), *Energy, Development and Women, Nurturing Nature,* Earthcare Books, Bombay].

Glossary

Abiotic: Non-living parts of the environment, e.g., air, water, coal, geological structures, temperature, topography etc.

Amniocentesis: Surgical procedure for obtaining a sample of amniotic fluid from the amniotic sac (in the uterus of a pregnant woman) by inserting a hollow needle through the abdominal wall; used in diagnosing certain genetic defects or possible obstetric complications.

Amniotic fluid: It is the nourishing and protecting liquid contained by the amnion of a pregnant woman.

Androcentric: Centred on, emphasizing or dominated by males or masculine interests: an androcentric society; an androcentric religion.

Anthropocentric ethic: The belief that only humans have value and that the environment solely exists for the benefit of humans; nature has no rights.

Anti-estrogenic: A substance that inhibits the physiological action of an estrogen.

Assarting: It is the act of clearing forested lands for use in agriculture or other purposes.

Bio-energy: Energy derived from bio-fuel.

Bio-ethics: A field of study concerned with the ethics and philosophical implications of certain biological and medical procedures, technologies and treatments like organ transplants, genetic engineering and care of the terminally ill.

Biologic diversity (species diversity, biotic diversity, bio-diversity): Usually equivalent to species richness, i.e., the number of different species existing in some location or under some condition (such as pollution). "Biologic diversity" may also be used in a more general sense to refer to the number of higher taxonomic levels or types and amounts of organismal relationships in some location or under some condition, e.g., the number of genera, families, orders or phyla present; or genetic pathways; or the number of biotic communities present; or the number of energy, nutrient, or food-chain pathways present. There are many measures and indices. Frequently used relationships are those for proportions of plants, animals or areas present.

Biomass (protoplasm mass): (1) The total weight (at a given time) of living organisms of one or more species per unit of space (species biomass) or of all the species in a community (community biomass). Phytomass expresses weight of plants only. (2) The total weight of matter incorporated into all organisms (living and dead).

Bio-region: An area defined by natural ecological systems such as a river watershed.

Bio-regionalism: Philosophies, ecological practices and politics built around the idea that a place's natural features and edges suggest the basis for understanding it and inhabiting it. Scientifically, this means connecting ecology to anthropology through geography: a seamless interdependency between ecosystem, culture and region. Most versions of bio-regionalism share the following areas of focus:

- Those who actually live in a bioregion best understand how to manage it. Top-down solutions from far away are suspect.
- Local democracy is based on direct participation and small-group discussion. (As Leopold Kohr put it, "If something is wrong, then something is too big.")
- Food is best grown and bought locally.
- Wellers begin to understand a place by reinhabiting it which means learning about its ecosystems, animals, water sources, weather, soil types, waste management, ecological strengths and traumas and resources for ecologically gentle living. Synchronising one's mood with these means learning to feel at home in a particular place.
- Developments that would damage the local environment – shopping malls, tract housing, factories etc. – should be firmly and consistently opposed. Locally made products are preferred over those shipped from a distance or made locally through mass production both of which transfer capital to outside sources.

- Respect for the rights, needs, customs, privacy, and knowledge of indigenous people living in the area.
- Living sustainably means ecologically sensible practices such as reuse and recycling, water and power conservation and reduction of trash and other wastes

Biosphere: That part of the earth's crust, waters, and surrounding atmosphere which is inhabited by living organisms.

Biota: The plant and animal life of an ecosystem; often referred to as flora and fauna.

Bio-technology: Biological manipulation of living organisms to produce foods, drugs and other products for humans.

Biotic: All natural living organisms in a planning area and their life processes. A resource classification category which sub-divides natural resources and properties into biotic or abiotic (non-living) entities and characteristics. Human activities are not usually grouped with either of these categories because they are not considered to be a part of a planning area's natural entities or characteristics. Nevertheless, native cultures have been actively influencing the course of biotic activity in some areas for centuries.

Cartesian: Of or pertaining to Descartes, his mathematical methods, or his philosophy, especially with regard to its emphasis on logical analysis and its mechanistic interpretation of physical nature.

Chromosome: The self-replicating genetic structure of cells containing cellular DNA that bears the linear array of genes in its nucleotide sequence.

Corporeality: Of the nature of the physical body; bodily, material; tangible: corporeal property.

Cybernetics: The study of human control functions and of mechanical and electronic systems designed to replace them, involving the application of statistical mechanics to communication engineering.

Cyborg: A person whose physiological functioning is aided by or dependent upon a mechanical or electronic device.

CGIAR: The Consultative Group on International Agricultural Research.

DDT: Dichlorodiphenyltrichloroethane: A chlorinated, hydrocarbon-based insecticide. It was originally considered safe and effective but proved to have negative human health and environmental consequences. It was banned in the US in 1972.

DNA (deoxyribonucleic acid): The molecule that encodes genetic information. DNA is a double-stranded molecule held together by weak bonds between base pairs of nucleotides. The four nucleotides in DNA contain the bases adenine (A), guanine (G), cytosine (C) and thymine (T). In nature, base pairs form only between A and T and between G and C; thus the base sequence of each single strand can be deduced from that of its partner.

Ecocentric: Term coined by Warwick Fox in preference to biocentric (environment-centred). Both contrast with the more human-focused perspective prevalent in industrialized nations.

Ecological economics: Non-traditional economics that focus attention not only on allocation and distribution of resources but also on the larger ecosystem or environment. The economy is viewed as a subsystem of the larger and finite ecosystem. Attention centers on the flow of matter and energy from the environment as raw materials and back to the environment as waste.

Ecology: The study of the relationship between all living organisms and the environment.

Ecological limit: The carrying capacity of a given area.

Ecosphere: Biosphere.

Ecosystem (bioc(o)enosis, biogeoc(o)enosis, biogeoc(o)enose):

1. A complete, interacting system or unit of organisms in a space considered together with their environment, e.g., a marsh, a watershed, a lake, etc., mentally isolated for purposes of study or management. A flow of energy leads to clearly defined food and feeding relationships, biologic diversity, and biogeochemical cycles (i.e., exchange of materials between living and nonliving parts) operating as an integrated system. "Ecosystem" is the preferred term in English while "biocoenosis" or "biogeocoenosis" is preferred by writers using or familiar with the Germanic and Slavic languages.
2. A nameable natural unit that includes not only the total array of plant and animal species in an environment, but also the energy to power the system.
3. A biotic community plus its abiotic environment.
4. Any complex of living organisms with its environment.
5. An interconnected community of plants and animals, including humans, and the physical environment within which they interact (USFS 1999).

Eco-politics: the study of the interrelation between political and ecological issues and problems.

Endometrial cells: Cells lining the uterus

Egocentric: Having or regarding the self or the individual as the centre of all things: an egocentric philosophy that ignores social causes; having little or no regard for interests, beliefs, or attitudes other than one's own; self-centered: an egocentric person; egocentric demands upon the time and patience of others.

Environmental ethics: Moral relations that hold between humans and the natural world.

Environmental refugees: Migrants from an area devastated by natural or technological hazards.

Esoterica: hings understood by or meant for a select few; recondite matters or items.

Ethics: A code of behaviour regarding what is appropriate and inappropriate.

Empiricism: the doctrine that all knowledge is derived from sense experience. Compare rationalism.

Epigenetic: For the unfolding of an organism or the theory that plants and animals develop in this way. In biology, the term epigenetics refers to heritable changes in phenotype (appearance) or gene expression caused by mechanisms other than changes in the underlying DNA sequence.

Eugenics: The study of or belief in the possibility of improving the qualities of the human species or a human population, esp. by such means as discouraging reproduction by persons having genetic defects or presumed to have inheritable undesirable traits (negative eugenics) or encouraging reproduction by persons presumed to have inheritable desirable traits (positive eugenics).

Factory farming: The mechanized, high-tech production of animals for human consumption. Growth hormones and antibiotics used to enhance growth of animals may pose health risks to humans.

FAO: Food and Agriculture Organization. A United Nations agency based in Rome and entrusted with the responsibility to improve efficiency of production and distribution of agricultural products.

GATT: General Agreement on Tariffs and Trade: A multilateral trade agreement regulating the world trade system.

Genetic diversity: The genetic variability within a group of related organisms.

Genetic engineering: The manipulation of genetic material for economic purposes.

Germ plasm: It is the protoplasm of the germ cells containing the chromosomes.

Food chain: A hierarchy of organisms, each feeding on the lower one.

Fossil fuels: Mineral fuels that occur in rock formations. They include coal, oil, and natural gas and provide a majority of the energy used in the world.

Global commons: Natural systems and resources that do not belong to any one country, e.g., the atmosphere and the oceans.

Gross National Product (GNP): The value of goods and services produced in a country plus remittances received from abroad.

Green politics: Politics centring on issues surrounding environmental problems.

Green Revolution: The effort organized by the United Nations in the 1960s to increase world food production by introducing high-yield varieties of rice, wheat, and maize and new techniques, including irrigation and use of pesticides.

Gross Domestic Product (GDP): The value of goods and services produced in the country.

Hermeneutics: Formal study of appropriate methods of interpretation [hermêneuma] first developed as a formal discipline of study by Schleiermacher. Following the work of Dilthey, Gadamer, and Ricouer, the hermeneutical process is often regarded as involving a complex interaction between the interpreting subject and the interpreted object. The task is complicated by the apparent circularity of understanding particular elements in light of the text as a whole, which can in turn be understood only by reference to them.

Imputation: The act of imputing or ascribing; attribution.

In vitro: Studies performed outside a living organism such as in a laboratory

Human ecology: An area of inquiry concerned with the relationship of humans and their activities with the physical environment.

Monoculture: An agricultural practice in which a single species of plant is cultivated in an area. It requires large amounts of fertilizers and pesticides.

Morphology: The branch of biology that deals with the form and structure of organisms without consideration of function. The form and structure of an organism or one of its parts: the morphology of a cell; the morphology of vertebrates.

Mutation rate: Is the chance of a mutation occurring in an organism or gene in each generation (or, in the case of multicellular organisms, cell division). The mutation frequency is the number of individuals in a population with a particular mutation, and tends to be reported more often as it is easier to measure (for instance, there is no need to restrict the population to experiencing only one generation, as needed to measure mutation rate). This is important in fields such as evolutionary biology and oncology.

Necrophilous: Having a preference for dead tissue. Used of certain bacteria.

Ocularcentrism: Amongst the senses, Plato gave primacy to sight. When he decided that we had five senses, Aristotle ranked sight over hearing: 'Of all the senses, trust only the sense of sight'. Plato and Aristotle closely associated vision and reason. This has been a persistent bias in Western culture. Thinking is associated with visual metaphors: 'observation' privileges visual data; phenomenon (Greek: 'exposing to sight'); definition (from definire, to draw a line around); insight, illuminate, shedding light, enlighten, vision, reflection, clarity, survey, perspective, point of view, overview, farsighted. Other words associated with thinking also have visual roots: intelligent, idea, theory, contemplate, speculate, bright, brilliant, dull. And there is no shortage of commonly used phrases which emphasize the primacy of the visual:

- Seeing is believing
- Let me see, I see
- I'll believe it when I see it with my own eyes
- Seeing eye to eye
- It's good to see you
- Love at first sight
- What does she see in him?
- In the mind's eye
- Draw your own conclusions
- See what I mean?

Oedipal complex: A complex of males; desire to possess the mother sexually and to exclude the father; said to be a source of personality disorders if unresolved [syn: Oedipus complex]

Omnivore: An animal that eats both plants and animals. Humans are good examples.

Ontology: The branch of metaphysics that studies the nature of existence or being as such.

Organic agriculture: The practice of growing crops without chemical fertilizers and pesticides, but otherwise similar to alternative agriculture.

Phallocentrism: A doctrine or belief centered on the phallus, esp. a belief in the superiority of the male sex.

Pederasty: Sexual relations between two males, esp. when one of them is a minor.

Polemical: A controversial argument, especially one refuting or attacking a specific opinion or doctrine. A person engaged in or inclined to controversy, argument, or refutation

Positivism: Belief that natural science, based on observation, comprises the whole of human knowledge. Positivists like Auguste Comte, then, reject as meaningless the claims of theology and metaphysics. The most influential twentieth-century version is logical positivism.

Proletariat: The word 'proletarius' in the ancient world meant those who were too poor to serve the state by property, and who served it instead by manufacturing labour power. Their role was to produce children; and since the historical burden of this task has fallen more on women than men, it is no mere modish gesture to claim that the proletariat is a woman.

Realism:

(1) Interest in or concern for the actual or real, as distinguished from the abstract, speculative, etc.
(2) The tendency to view or represent things as they really are.

Sierra Club: An environmental organization founded by John Muir in 1892 that promotes public education, litigation and outings and conferences.

Spermatozoon: One of the minute, usually actively motile gametes in semen, which serve to fertilize the ovum; a mature male reproductive cell.

Social ecology: A radical ecological position championed by Murray Bookchin maintaining that environmental problems can be traced to a hierarchical political/economic system that dominates humans and nature. It calls for a decentralized, democratic, and agricultural-based society.

Stockholm Conference: The 1972 United Nations Conference on the Human Environment. Many scholars assert that this conference was the springboard for the modern environmental movement.

Sustainable development: The perspective emphasizing the need to reconcile present and future economic needs through environmental management.

Third World: Less developed countries that have low per capita incomes, large agricultural sectors, and a shortage of most kinds of capital.

Topsoil: The top layer of the soil that contains large amounts of organic matter. It is often viewed as a transitional region between the living and nonliving segments of the larger biosphere.

Transgenic: of, pertaining to, or containing a gene or genes transferred from another species: transgenic mice.

Unequal exchange: A pattern describing trade relations between two or more countries when one country benefits more than another.

Value added: A means of increasing the value of agricultural commodities by improvements (e.g., breeding wheat with high protein content) or processing (e.g., grinding wheat into flour).

World Wildlife Fund: An organization founded in the US in 1961 and working globally to protect endangered wildlife and wetlands.

Water table: The level below the surface at which the ground is saturated with water.

Wetlands: Several different types of habitats containing water, e.g., marshes, swamps and bogs. They act as filters for rivers and streams and minimize the effects of flooding by storing water.

Bibliography

Adams, Carol J. (1991) "Ecofeminism and Eating of Animals", *Hypatia*, Spring, Vol. 6, No. 1, pp. 125–145.

Agarwal, Bina. (1985) "Women, Poverty and Agricultural Growth in India", *The Journal of Peasant Studies*, Vol. 13, No. 4.

——. (1986) *Cold Hearths and Barren Slopes: The Woodfuel Crisis in the Third World*. Allied Publishers, New Delhi.

——. (1992) "The Gender and Environment Debate: Lessons from India", *Feminist Studies*, Vol. 18, No. 1, pp. 119–157.

——. (1997) "A Field of One's Own: Gender and Land Rights in South-Asia", in *A Just Right*, Nitya Rao and Luise Rürup (eds.), Friedrich Ebert Stiftung, New Delhi, pp. 37–43.

Anandi Team. (No Date) "Understanding Women's Experiences", *Natural Resource Management*, Agakhan Foundation.

Andharia, Janki and Chandan Sengupta. (1998) "Environmental Movement: Global Issues and the Indian Reality", in *Towards People-Centered Development*, Part II, M. Desai, A. Monteiro and L. Narayan (eds.), Tata Institute of Social Sciences, Mumbai, pp. 422–449.

Aurora, G.S. (1991) "Green Revolution Technologies and Dryland Agriculture", *Sociological Bulletin*, (March-September), Vol. 40 (1and 2), p. 119.

Bakshi, Rajni. (2007) *An Economics for Well-Being*. Centre for Education and Documentation (CED), Mumbai.

Barot, Nafisa. (1997) "A People's Movement towards Creating Sustainable Drinking Water Systems in Rural Gujarat", in *A Just Right*, Nitya Rao and Luise Rürup (eds.), Friedrich Ebert Stiftung, New Delhi, pp. 470–486.

Baviskar, Amita. (1997) "Women and Forests: Dependence without Control", in *A Just Right*, Nitya Rao and Luise Rürup (eds.), Friedrich Ebert Stiftung, New Delhi, pp. 347–352.

Bennholdt-Thomsen, Veronika, Claudia von Werlhof and Nicholas Faraclas. (2001) *There is an Alternative: Subsistence and Worldwide Resistance to Corporate Globalization*. Zed Books, UK, USA, Australia.

Boulding, Elise. (1976) *The Underside of History: A View of Women through Time*. Boulder, Colorado, Westview Press.

Brahme, Sulabha. (1997) "New Economic Policy, Land Rights and Women's Question with Reference to Maharashtra", in *A Just Right*, Nitya Rao and Luise Rürup (eds.), Friedrich Ebert Stiftung, New Delhi, pp. 198–220.

Braidotti, Rosi, Eva Charkiewicz, Sabine Hausler and Saskia Wieringa. (1994) *Women, the Environment and Sustainable Development: Towards a Theoretical Synthesis*. Zed Books in association with INSTRAW, London.

Carson, Rachel. (1970) *Silent Spring* (first published in 1962), Houghton-Mifflin, Boston.

Chattopadhyaya, Debiprasad. (1959) *Lokayata: A Study in Ancient Indian Materialism*. People's Publishing House, New Delhi.

Cochrane, Regina. "The Sustainability of Tradition? From the Subsistence Perspective to a Feminist Ecoregionalism", University of Calgary, Canada, from the website: http://www.arts.ualberta.ca/cms/cochrane.pdf. (as per 14-5-09).

Corea, Gena, Jalna Hanmer, Renate D. Klein, Janice G. Raymond and Robin Rowland. (1987) "Prologue", in *Made to Order*, Patricia Spallone and Deborah Lynn Steinberg (eds.), The Athene Series, Pergamon Press, Oxford, New York, Bejing, Sao Paulo, Sydney.

Daly, H.E. and J.B. Cobb Jr. (1990) *For the Common Good*. Green Print, USA.

Dasmann, R.F. (1988) "Towards a Biosphere Consciousness", in *The Ends of the Earth: Perspectives on Modern Environmental History*, D. Worster (ed.), Cambridge University Press, Cambridge.

Datar, Chhaya. (2007) *Widows' Woes*. Unpublished Report, TISS, Mumbai.

Datye, K.R. (1997) "Banking on Biomass: A New Strategy for Sustainable Prosperity based on Renewable Energy and

Dispersed Industrialisation", assisted by Suhas Paranjape and K.J. Joy, Centre for Environmental Education, Ahmedabad.

Dhere, R.C. (1978) *Lajjagouri*. Shreevidya Prakashan, Pune, India.

Diamond, I. and G.F. Orenstein (eds.) (1990) *Reweaving the World: The Emergence of Ecofeminism*. Sierra Club Books, San Francisco, pp. ix–xv.

Diprose, Rosalyn. (1995) "Bisemiotics and Ethics", in *Biopolitics*, Vandana Shiva and Igunn Moser (eds.), Zed Press, New Jersey and London, pp. 162–174.

Duraiswamy, S. (2009) "Philosophy of Biology", in *History of Science, Philosophy and Culture in Indian Civilisation*, Vol. XII, Part 6 (Life and Organicism), N.S. Rangaswamy (ed.), D.P. Chattopadhyaya (Gen. ed.), Centre for Studies in Civilisations, New Delhi.

Eckersley, R. (1992) *Environmentalism and Political Theory: Towards an Ecocentric Approach*. UCL Press, London.

Eisler, Riane. (1987) *The Chalice and the Blade: Our History, Our Future*. Harper & Row, San Francisco.

——. (1995) "Messages from the Past: The World of the Goddess", in *This Sacred Earth*, Roger S. Gottlieb (ed.), Routledge, New York, pp. 369–379.

Engels, Frederick. (1973) *Condition of the Working Class in England: From Personal Observation and Authentic Sources*. Progress Publishers, Moscow.

——. *Dialectics of Nature*. (Unfinished work)

Fowler, Cary. (1995) "Biotechnology, Patents and the Third World", in *Biopolitics*, Vandana Shiva and Ingunn Moser (eds.), Zed Press, New Jersey and London, pp. 214–225.

Gadgil, Madhav and Ramchandra Guha. (1993) *This Fissured Land: An Ecological History of India*. UCL Press, Berkeley, Los Angeles.

——. (1995) "Passing on the Costs Islands of Prosperity, Oceans of Poverty", in *Equity and Power* (UNRISED sponsored), Routledge, London, pp. 34–60.

——. (1998) "Towards a Perspective on Environmental Movement in India", in *Towards People-centered Development*, Part II, Tata Institute of Social Sciences, Mumbai, pp. 448–472.

Gadgil, S.R. (1998) *Lokayat*. People's Publishing House, Mumbai.

Ganesh, Kamala. (2009) "In Search of the Great Indian Goddesses: Motherhood Unbound", in *Motherhood in India*, Maithreyi Krishnaraj (ed.), Routledge, New Delhi.

Goering, P., H. Norberg-Hodge and J. Page. (1993) *From the Ground Up*. Zed Books, London.

Goodin, R.E. (1992) *Green Political Theory*. Polity Press, Cambridge, UK, pp. 19–83.

Goodland, R. and Herman Daly. (1992) "The Reasons Why Northern Income Growth is not the Solution to Southern Poverty", *International Journal for Sustainable Development*, Vol. 1, No. 2, pp. 23–30.

Gopal, K.S. and M. Shashikumar. (1993) Public Distribution System, Towards a New Paradigm by Centre for Environmental Concerns, Hyderabad (unpublished), p. 5.

Government of India (1974) "Towards Equality", a Report of the Committee on the Status of Women in India.

Grosz, E. (1989) "Luce Irigaray and Sexual Difference", in *Sexual Subversions: Three French Feminists*, Allen & Unwin, Sydney, pp. 101–183.

Guha, Ramchandra and J. Martinez-Alier. (1998) "The Merchandising of Biodiversity", in *Varieties of Environmentalism: Essays North and South*, Oxford University Press, New Delhi, pp. 109–120.

Haraway, Donna J. (1991) *Simians, Cyborgs and Women: The Reinvention of Nature*. Free Association Books, London.

Harding, Sandra. (1986) *The Science Question in Feminism*. Cornell University Press, Ithaca and London.

——. (1987) "Epistemological Questions: Conclusion", in *Feminism and Methodology*, ed. by self, Indiana University Press, Bloomington and Indianapolis.

——. (1987) "Introduction: Is There a Feminist Method?", in *Feminism and Methodology*, ed. by self, Indiana University Press, Bloomington and Indianapolis.

——. (1998) *Is Science Multi-Cultural?*, Indiana University Press, Bloomington and Indianapolis.

Harford, Barbara and Sarah Hopkins (eds.) (1984, reprint 1985) *Greenham Common: Women at the Wire*. The Women's Press, London.

Harriss-White, Barbara. (2007) "Commercialisation, Commodification and Gender Relations in Post-Harvest Systems in Rice in South Asia", in *Gender, Food Security and Rural Livelihoods*, Maithreyi Krishnaraj (ed.), Stree, Kolkatta.

Hartsock, Nancy C.M. (1983) "The Feminist Standpoint: Developing the Ground for a Specifically Feminist Historical Materialism", in *Money, Sex and Power: Towards a Feminist Historical Materialism*, Northeastern University Press, Boston, pp. 258–272.

——. (1987) "The Feminist Standpoint: Developing the Ground for a Specially Feminist Historical Materialism", in *Feminism and Methodology*, Sandra Harding (ed.), Indiana University Press, Bloomington and Indianapolis.

Henderson, Hazel. (1981) *The Politics of the Solar Age: Alternatives to Economics*. Anchor Press/Doubleday, New York.

Hoffmeyer, Jesper. (1995) "Biosemiotics and Ethics", in *Biopolitics*, Vandana Shiva and Igunn Moser (eds.), Zed Books, London and New Jersey, pp. 141–161.

Hubbard, Ruth. (1995) "Human Nature", in *Biopolitics*, Vandana Shiva and Igunn Moser (eds.), Zed Books, London and New Jersey, pp. 38–51.

Huges, J. (2005) "Green or Environmental Movements", in *Berkshire Encyclopedia of World History*, by William McNeill, Jerry Bently and David Christian, pp. 866–870.

Illich, I. (1973) *Tools for Conviviality*. Fontana/Collins, London.

Jackson, Cecile. (1995) "Radical Environmental Myths: A Gender Perspective", *New Left Review*, March–April, No. 210, pp. 124–127.

Jain, Shobhita. (1984) "Women and People's Ecological Movement: A Case Study of Women's Role in the Chipko Movement in Uttar Pradesh", *EPW*, Vol. XIX, No. 41, pp. 1788–1794.

Jansen, Sue Curry. (1990) "Is Science a Man? New Feminist Epistemologies and Reconstruction of Knowledge", *Theory and Society*, Vol. 19, pp. 235–246.

Jeffery, Roger, Patricia Jeffery and Andrew Lynn. (1989) "Taking Dung Work Seriously: Women's Work and Rural Development in North India", *EPW*, April, pp. WS32-WS37.

Jodha, N.S. (1986) "Common Property Resources and Rural Poor in Dry Regions of India", *EPW*, Vol. XXI, No. 27, July 5.

Jones, Alwyn K. (1990) "Social Symbiosis: A Gaian Critique of Contemporary Social Theory", *The Ecologist*, Vol. 20, No. 3, May/June.

Joshi, Divya. (2003) *Gandhiji on Environment*. Mani Bhavan Gandhi Sangrahalaya, Mumbai.

Kabeer, Naila. (1994) *Reversed Realities: Gender Hierarchies in Development Thought*. Verso, London and New York.

Keller, Evelyn Fox. (1985) *Reflections on Gender and Science*. Yale University Press, New Haven and London.

——. (1989) "The Gender/Science System: or, Is Sex to Gender as Nature to Science", in *Feminism and Science*, Nancy Tuana (ed.), Indiana University Press, Bloomington and Indianapolis, pp. 33–44.

Keown, Damien (1996) *Buddhism: A Very Short Introduction*. OUP (Indian Edition), New York.

Khanna, Madhu. (1995) "Nature as Feminine: Ancient Vision of Geopiety and Goddess Ecology", *Prakriti: The Integral Vision*, Vol. 5, *Man in Nature*, Indira Gandhi National Centre for the Arts, New Delhi.

King, Roger J.H. (1991) "Caring About Nature: Feminist Ethics and the Environment", *Hypatia*, Spring, Vol. 6, No. 1, pp. 75–89.

Kollek, Regine. (1995) "The Limits of Experimental Knowledge: A Feminist Perspective on the Ecological Risks of Genetic Engineering", in *Bioplitics*, Vandana Shiva and Ingunn Moser (eds.), Zed Books, London and New Jersey, pp. 95–111.

Kosambi, D.D. (2009) "At the Crossroads: A Study of Mother-Goddess Cult Sites", in *Combined Methods in Indology and Other Writings*, Brajadulal Chattopadhyaya (ed.), Oxford University Press, New Delhi, (Second Edition), pp. 830–859.

Krishna, Sumi. (2004) *Livelihood and Gender: Equity in Community Resource Management*. Sage Publications, New Delhi.

Krishnaraj, Maithreyi. (1988) "Women's Studies: Case for a New Paradigm", *EPW*, April 30, pp. 892–894.

——. (2007) "Food Security, Agrarian Crisis and Rural Livelihoods: Implications for Women", in *Gender, Food Security and Rural Livelihoods*, ed. by self, Stree, Kolkatta, pp. 24–72.

——. (2009) "Motherhoods, Mothers and Mothering: A Multidimensional Perspective", in *Ideology of Motherhood,* Taylor & Francis, Routledge, New Delhi, pp. 9–43.

Krishnaraj, Maithreyi and Aruna Kanchi. (2008) *Women Farmers of India*. National Book Trust of India, New Delhi.

Krishnaswamy, S. (1988) "Female Infanticide in Contemporary India: A Case Study of the Kallars of Tamil Nadu", in *Women in Indian Society: A Reader,* Rehana Ghadially (ed.), Sage Publications, New Delhi.

Kulkarni, Sharad. (1997) "Natural Resources, New Economic Policy and Marginalized Communities", in *A Just Right,* Nitya Rao and Luise Rürup (eds.), Friedrich Ebert Stiftung, New Delhi, pp. 337–346.

Kumar, Gopa R. (1997) "Land Reforms and its Impact on Poor: An Interstate Analysis", in *A Just Right,* Nitya Rao and Luise Rürup (eds.), Friedrich Ebert Stiftung, New Delhi, pp. 122–131.

Kuhn, Thomas S. (2000) *The Road Since Structure,* University of Chicago Press, Chicago.

Kurien, C.T. (1992) *Growth and Justice: Aspects of India's Development Experience*. Oxford University Press, Bombay, pp. 270–77.

——. (1978) *Poverty, Planning and Social Transformation*. Allied Publishers, Bombay and Calcutta.

Lerner, Gerda. (1986) *The Creation of Patriarchy*. Oxford University Press, New York.

Lloyd, Christopher. (1986) *Explanation in Social History*. Basil Blackwell, UK.

McLaughlin, A. (1993) *Regarding Nature: Industrialism and Deep Ecology*. SUNY, New York.

McCormick, John. (1989) *The Global Environmental Movement: Reclaiming Paradise*. Belhaven Press, London, UK.

Meillasoux, C. (1981) *Maidens, Meal and Money: Capitalism and Domestic Community*. Cambridge University Press, London.

Menon, Ajit et al. (2007) "Hivre Bazar: A Model Watershed Experiment?", in *Community-based Natural Resource Management: Issues and Cases from South Asia,* Sage Publications, New Delhi, p. 28.

Menon, Gayatri. (1996) 'Re-negotiating Gender: Enabling Women to Claim their Right to Land Resources', paper presented at the special event: Women's Legal Rights to Land/Property, held at the NGO Forum (Habitat II). Organized by AWAS & COHR, in partnership with UNCHS, UNEP & UNIFEM.

Menon, Meena. (2005) *Organic Cotton Cultivation: Re-inventing the Wheel*. Deccan Development Society, Hyderabad and Kalpavriksh, Pune.

Merchant, Carolyn. (1983) *The Death of Nature: Women, Ecology and the Scientific Revolution*. Harper and Row, San Francisco.

——. (1995) *Earthcare: Women and the Environment*. Routledge, New York, pp. 209–224.

Mies, Maria. (1986) *Patriarchy and Accumulation on a World Scale*. Zed Books, London.

——. (1988) "Social Origin of the Sexual Division of Labour", in *Women, the Last Colony*, Kali for Women, New Delhi, pp. 65–86.

——. (1996) 'Women, Food and Global Trade', a paper published by Institute Für Theorie and Praxis der Subsistenz, Bielefeld, Germany.

——. (1997) "Do We Need a New Moral Economy?", *Canadian Women's Studies*, Vol. 17, No. 2, University of York.

Mies, Maria and Vandana Shiva. (1993) *Ecofeminism*. Zed Books, London; Fernwood, Halifax, Nova Scotia; Kali for Women, New Delhi.

Mies, Maria and Veronika Bennholdt-Thomsen. (1999) *The Subsistence Perspective: Beyond the Globalised Economy*. Zed Books, London and New York.

Mies, Maria, Claudia Von Werlhof and Veronika Bennholdt-Thomsen. (1988) *Women, the Last Colony*. Kali for Women, New Delhi.

Miller, Barbara. (1985) "The Unwanted Girls: A Study of Infant Mortality Rates", *Manushi: Journal of Women and Society*, No. 29, pp. 52–54.

Mills, Patricia Jagentowicz. (1991) "Feminism and Ecology: On the Domination of Nature", *Hypatia*, Spring, Vol. 6, No. 1, pp. 162–177.

Molyneux, Maxine and Deborah Lynn Steinberg. (1995) "Mies and Shiva's Ecofeminism: A New Testament?", *Feminist Review*, Spring, No. 49, pp. 86–107.

Okasha, Samir. (2002, 2006) *Philosophy of Science: A Very Short Introduction*. Oxford University Press, New York, pp. 120–149.

Omvedt, Gail and Govind Kelkar. (1993) "Gender and Technology: A Critical Perspective (unpublished)", *Gender and Development Studies*, School of Environment, Resources and Development, Bangkok, Asian Institute of Technology.

Parasuraman, S. (1993) Impact of Displacement by Development Projects on Women in India. ISS: The Hague. ISS working paper series/number: 159).

Patnaik, Utsa. (1996) "Globalisation and Agriculture: Export-oriented Agriculture and Food Security in Developing Countries and India", *EPW*, September, Vol. XXXI, Special Issue Nos. 35-37, pp. 2429–2449.

——. (1998) "Scenario in Developing Countries and India: Export-oriented Agriculture and Food Security", in *The Great Grain Drain*, Books for Change, Bangalore.

Patnaik, Utsa in collaboration with Subhasini Ali. (1993) 'Likely Impact of Economic Liberalisation and Structural Adjustment on the Food Security System in India', paper presented at ILO/National Commission for Women, National Workshop on Employment Equality and Impact of Economic Reform for Women, New Delhi, India.

Plumwood, Val. (1991) "Nature, Self and Gender: Feminism, Environmental Philosophy and the Critique of Rationalism", *Hypatia*, Spring, Vol. 6, No. 1, pp. 3–27.

——. (1992) "Feminism and Ecofeminism", *Ecologist*, Vol. 22, No. 1, January/February, UK, pp. 8–13.

——. (1993) *Feminism and Mastery of Nature*. Routledge, London and New York, pp. 104–119.

Polanyi, Karl. (1944, 1997) *The Great Transformation: The Political and Economic Origins of Our Time*. Beacon Press, Boston, USA.

Radcliff, M. (1987) *Sustainable Development: Exploring the Contradictions*. Routledge, London.

Rege, M.P. (1995) Samkhya Theory of Matter, Prakrati, The Nature of Matter, Vol. 4, Indira Gandhi National Centre for Arts, New Delhi.

Rose, Hillary. (1988) "Beyond Masculinist Realities: A Feminist Epistemology for the Sciences", in *Feminist Approaches to Science*,

Ruth Blier (ed.), Pergamon Press, USA, UK, China, Germany, Brazil.

——. (1994) *Love, Power and Knowledge*. Polity Press in association with Blackwell Publishers, Cambridge, UK.

Sallah, Ariel. (1991) Book Review. "Staying Alive: Women, Ecology and Development", by Vandana Shiva, *Hypatia*, Spring, Vol. 6, No. 1.

——. (1993) "Class, Race and Gender Discourse in Ecofeminism and Deep Ecology Debate", *Environmental Ethics*, Fall, Vol. 15, No. 3, pp. 225–244.

Salunkhe, Vilasrao, Vasudha Lokur and Ganesh Pangare. (1997) "Pani Panchayat Model of Water Management", in *A Just Right*, Nitya Rao and Luise Rürup (eds.), Friedrich Ebert Stiftung, New Delhi, pp. 441–456.

Sarin, Madhu. (1996) *Joint Forest Management: The Haryana Experience*. Centre for Environment Education, Ahmedabad.

——. (1997) "Gender and Equity Concerns in Joint Forest Management", in *A Just Right*, Nitya Rao and Luise Rürup (eds.), Friedrich Ebert Stiftung, New Delhi, pp. 267–336.

——. (1998) "Community Forest Management: Whose Participation?", in *The Myth of Community-Gender Issues in Participatory Development*, Irene Guijt and Meera Kaul Shah (eds.), Intermediate Technology Publications, London, UK, pp. 121–130.

Sarin, Madhu and Renu Khanna. (1993) "Women Organize for Wasteland Development: A Case Study of SARTHI in Gujarat", in *Wasteland Development in India*, Neera Burra and Andrea M. Singh (eds.), Sage Publications, New Delhi, London, pp. 129–169.

Sarkar, Saral. (1983) Review Article. *Marxism and Productive Forces: A Critique*, A book by Otto Uelrich in der Saekgarse des industrie systems, Berlin Rotbuch Veriag (1979).

Sen, Amartya and Sunil Sengupta. (1983) "Malnutrition of Rural Children and the Sex Bias", *EPW*, Vol. XVIII, Nos. 19-21, pp. 855–864.

Sen, Gita. (1985b) 'Changing International Perspectives towards Women and Food: An Appraisal', paper presented at the international workshop on Women's Role in Food, Self-sufficiency and Food Strategies, ORSTOM/CIE, Paris.

Sen, Gita and Caren Grown. (1985a) DAWN, Development Alternatives with Women for a New Era. *Development, Crises and*

Alternative Visions: The Third World Women's Perspectives. DAWN Secretariate at ISST, New Delhi, pp. 43–67, 71–82.

Shearer, Rhonda Roland. (1997) Book Review. "Ecofeminism", by Mies and Shiva, *Signs*, Winter, pp. 496–501.

Shensu, Meena and Vasant Bhosale. (1990) *Imprisoning Womanhood: Report on Death and Desertion of Women in Sangli District*.

Shiva, Vandana. (1988) *Staying Alive*. Kali for Women, New Delhi.

——. (1989) *The Violence of the Green Revolution: Ecological Degradation and Political Conflict in Punjab*. Natraj Publisher, Dehradun, India.

——. (1991) *Ecology and Politics of Survival*. Sage Publications, New Delhi and The United Nations University Press, Tokyo, Japan.

——. (1993) *Monocultures of the Mind: Perspectives on Biodiversity and Biotechnology*. Third World Network, Malaysia.

——. (1995) "Biotechnological Development and the Conservation of Nature", in *Biopolitics: A Feminist and Ecological Reader on Biotechnology*, Vandana Shiva and Ingunn Moser (eds.), Zed Books, London, New Jersey and Third World Network, Penang, Malaysia, pp. 193–213.

Slater, Phil. (1979) *Origin and Significance of the Frankfurt School*, Routledge and Kegan Paul, London.

Spretnak, Charlene. (1990) "Ecofeminism: Our Roots and Flowering", in *Reweaving the World*, Irene Diamond and Gloria Femon Orenstein (eds.), Sierra Club Books, San Francisco, pp. 3–14.

Trainer, Ted. (1989) *Developed to Death: Rethinking Third World Development*. Green Prints, London, pp. 42–211.

Wagle, Subodh. (1998) "Energy, Development and Women in Nurturing Nature", in *Women at the Centre of Natural and Social Regeneration*, Chhaya Datar (ed.), Earthcare Books, Bombay, p. 29.

Wallerstein, I. (1974) "The Rise and Future Demise of the World Capitalist System", *Comparative Studies in Society and History*, Vol. XVI, No. 4, pp. 387–415.

Waring, Marilyne. (1988) *If Women Counted*. Harper and Row, San Francisco.

Weizsacker, Christine von. (1995) "Error-Friendliness and the Evolutionary Impact", *Biopolitics*, Zed Press, London, New Jersey, pp. 112–121.

Werlhof, Claudia (1988) "On the Concept of Nature and Society in Capitalism", in *Women: The Last Colony*, Maria Mies et al., Kali for Women, New Delhi.

Wood, John R. (2007) *The Politics of Water Resource Development in India*. Sage Publications, New Delhi.

World Bank Report. (2006) *India: Unlocking Opportunities for Forest Dependent People in India*. Report No. 34481-IN, Main Report: Volume 1, February 6, 2006, Agriculture and Rural Development Sector Unit, South Asia Region.

Index